Holley Carburetors & Manifolds
by Mike Urich & Bill Fisher

Table of Contents

Publisher:
   Bill Fisher

Editors:
   Bill Fisher, Carl Shipman

Cover design:
   Josh Young

Book design & assembly:
   Nancy Fisher, Chris Crosson

Text & caption typography:
   Marcia Redding, Lou Duerr

Drawings:
   Erwin Acuntius

Cover photos:
   Photomation

Photography:
   Bill Fisher, Mike Urich,
   Howard Fisher, others

Front cover shows a formidable racing combination—Holley's Double-Pumper 4150 Carburetor on a Holley Strip Dominator Manifold. The back cover is the compact Holley Model 4360 four-barrel on a Holley Street Dominator Manifold—an ideal street or RV combination.

An independent publication—not published by or associated with Holley Carburetor Division, Colt Industries Operating Corp.

ISBN: 0-912656-48-4
Library of Congress Catalog
      Card No. 76-5998
H. P. Book No. 48
© 1976 Printed in U.S.A. 6-76
H. P. Books, P. O. Box 5367
Tucson, AZ 85703
602/888-2150

Introduction

Why a book on Holley carburetors?

First—until May, 1972 when we printed the first edition of this book, no such book was available to help the automotive enthusiast who needed accurate and tested information on using Holley high-performance carburetors. Internal-combustion-engine textbooks have carburetion sections and entire books are devoted to carburetor design. But directing an automotive enthusiast to one of these books would not really help because it's all too easy to become bogged down in the theory and mathematics of what's happening. Some easy-to-understand illustrated descriptions of the carburetor and its workings are in automobile shop manuals, but this information is oriented toward *repairs* on a *normal* car—not one being tuned for *economy, high-performance* or *racing.*

Second—a lot of good information has been developed in Holley Engineering. But, there was no easy way to tell about it without a basic understanding of carburetion and the relation of the various systems in a carburetor—and how they work. It was a case of not being able to tell important details without complete background data. Details by themselves would get some users into difficulties because the relationships are not obvious.

Third—Holleys are the most widely used high-performance carburetors in the world, but no one piece of literature described the important ones in detail and told how to get the most out of them.

Finally—the past 14 years have seen a continuously increasing emphasis on emission controls. Because the carburetor is one of the main controls, understanding its relationship to other components in the systems has become essential. Everyone needs to know more about how these systems work—and the role the carburetor plays in emission reduction.

Photos and illustrations have been used freely to describe construction features and operation of these carburetors. Some

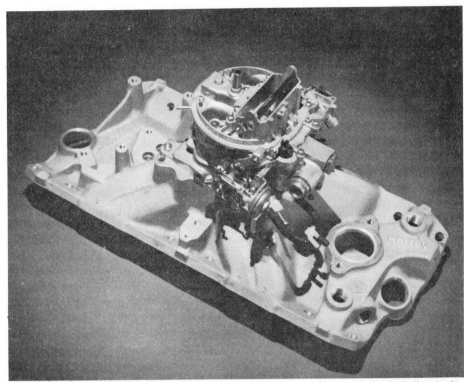

Holley Model 4360 four-barrel carburetor on a Street Dominator manifold. Holley is the only manufacturer supplying the entire carburetion system with all components carefully engineered to work together perfectly. They make manifolds, carburetors, fuel lines, fuel filters and fuel pumps—everything needed for feeding fuel to the engine.

show how special parts can be used to provide improved performance in specific applications . . . and how the stock parts should be applied or installed.

Because factory service and overhaul manuals are often unavailable, co-author Mike Urich provided how-to photo sequences on disassembly and assembly of the most commonly used Holley Carburetors. If you are using a late-model car on the road, be sure to observe the idle-setting information on the sticker in the engine compartment.

You will find a lot of tips on high-performance carburetors. We incorporated answers to all of the questions enthusiasts ask again and again at technical sessions, at the race track and at the drags. We have tried to dispel some of the rumors, myths and half-truths which have become part of the romance of using Holley carburetors.

As you might imagine, this book was not created overnight. In fact, Holley engineers and H. P. Books planned and worked together for seven years prior to printing the first edition. During that time, a lot of information was collected, photos made, drawings produced, and so forth. Holley continuously improves all of their carburetors: Stock replacement, high-performance and economy carburetors.

Any book of this type has a certain degree of obsolescence built into it because there is no way to capture more than a snapshot of development and the availability of parts at that final moment when the printing presses are turned on. So, it's up to you to keep track of what is happening in high-performance carburetion by keeping tuned to the availability of new parts and pieces from Holley.

Holley History

George Holley in a replica of his 1897 three-wheeler. He built it at age 19!

From its very beginnings, Holley Carburetor Company combined an intense interest in auto racing with a dedication to engineering excellence. Founded in 1902 by the Holley Brothers of Bradford, Pennsylvania, the company as we know it 75 years later grew out of their experiments with the infant horseless carriage.

At the age of 19, George M. Holley designed and built his first car—a single-cylinder three-wheeler capable of 30 miles per hour.

Fascination with speed and things mechanical soon led young Holley into motorcycle racing at which he excelled in national competition. Together with his brother Earl, they formed Holley Brothers to build motorcycle engines when they were not racing. Later, they built entire motorcycles.

This unique combination of talents led to still another original vehicle—long since disappeared from the auto scene—the Holley *Motorette*. Introduced in 1903, this jaunty little candy-apple-red sports model sold for $550—fully equipped. More than 600 of the 5.5-HP vehicles were built over a three-year period. Only three of these original cars are still alive today—one in the Holley lobby in Warren, Michigan.

As the fledgling auto industry was taking shape, the first hint of industrial specialization began to emerge. Sens-

ing this trend, the Holley brothers elected to concentrate on designing and building carburetors and ignition-system components for car makers such as Pierce-Arrow, Winton, Buick, and Ford. They left building basic vehicles to their customers and became original-equipment suppliers.

Their first original carburetor, called the "iron pot," appeared on the curved-dash Olds in 1904.

Today, Holley carburetors can be found on Chrysler, Ford, General Motors, International Harvester, White, Mack, Diamond Reo and other vehicles as well as some very exotic specialized applications such as the Corvette LT-1 and the Camaro Z-28. Holley ignition distributors have been standard equipment on thousands of vehicles built by International Harvester, Ford and others.

Should you want even more fuel flow or decide to alter your basic induction system, Holley makes hundreds of different carburetors for replacement applications. Holley carburetors have long been the front runners in the performance area: The original three-barrel, the NASCAR 4500, and more recently, a family of large two-barrels, "double-pumpers," individual-runner type carburetors for all-out racing and the Model 4165 small/large spread-bore for good emissions with performance.

In 1976, economy carburetors and a new line of *Dominator* Street and Strip Manifolds joined the Holley Induction Team. Holley is the *only* carburetor manufacturer in the world making the entire system: Fuel pumps, fuel lines, fuel filters, carburetors and intake manifolds.

In addition, high-performance fuel pumps and ignition kits have joined this family, along with valve covers and electrical components.

As you can see, Holley has always been closely identified with high performance. It was the cornerstone on which the company was built and continues as an important part of Holley Carburetor Division of Colt Industries.

Model NH Holley fed fuel to countless Model T's on which it was an original-equipment item.

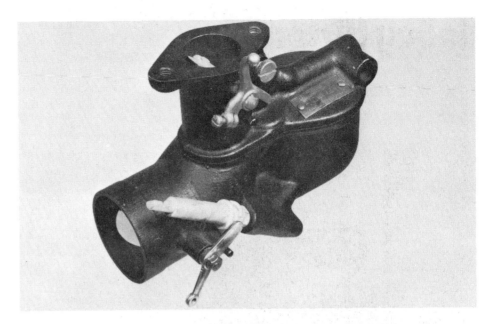

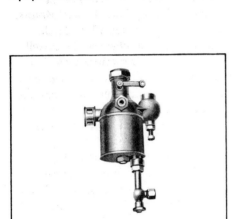

"Iron Pot" carburetor made by Holley Brothers for early curved-dash Oldsmobile in 1904 started their specialization into the carburetor business. Early customers included Buick, Pierce-Arrow and Winton.

$100 per horsepower! Holley "Motorette" Runabout cost $550 with 5-1/2 HP engine. 64-inch-wheelbase car weighed 600 pounds, featured sight-feed lubrication, planetary-drive transmission and a tilting steering wheel with lock. Carburetion? Would you have guessed a single-barrel Holley? 600 were made in 1903-04-05.

1926 Western Auto catalog advertised a new Holley at a really low price! They're worth more than that today. The carburetor is a Model NH for the Model T.

Engine Requirements

Whether you work on engines to make them run quicker, faster—or merely "sweeter," you'll get more performance after you've fully digested this chapter and the ones which follow on How Your Carburetor Works and Selecting and Installing Your Carburetor.

You don't need to worry about tough mathematical formulas because we've made the whole idea of carburetion simple.

AIR FLOW REQUIREMENTS

Because the air an engine consumes has to come in through the carburetor, knowing how much air the engine *can* consume will help you select the correct carburetor size.

How big should the carburetor be? Like the man said, "*It depends.*" It depends on:

Engine Displacement—expressed in cubic inches displacement (CID or cu. in.) or cubic centimeters (cc).

Type of Engine—two-cycle or four-cycle. Two-cycle engines have an intake stroke per cylinder every revolution. Four-cycle engines have an intake stroke per cylinder every other revolution.

Maximum RPM—the peak RPM which the engine will "see." In this area you must be absolutely honest and realistic without kidding yourself. "Dream" figures will only lead you into the trap of getting a too-big carburetor, causing problems discussed throughout this book.

Once you know the engine type, RPM and displacement, figuring the air-flow requirement in cubic feet per minute (CFM) gets real easy if we assume the engine has perfect "breathing" or 100% volumetric efficiency.

For 2-Cycle Engines

$$\frac{CID \times RPM \times Volumetric\ Efficiency}{1728} = CFM$$

For 4-Cycle Engines

$$\frac{CID}{2} \times \frac{RPM}{1728} \times Volumetric\ Efficiency = CFM$$

Example: 350 cubic inch engine

8000 RPM maximum

Assume volumetric efficiency of 1 (100%)

$$\frac{350\ CID}{2} \times \frac{8000\ RPM}{1728} \times 1 = 811\ CFM$$

A volumetric efficiency of 100% or 1 is not usually attainable with a naturally-aspirated (unsupercharged) engine. Thus, the engine in question will not flow 811 CFM of air (at standard temperature and pressure). Let's talk about volumetric efficiency so we can see how it affects the air-flow requirement.

VOLUMETRIC EFFICIENCY

Volumetric efficiency (V.E., η) indicates how well the engine breathes. The better the "breathing ability"—the higher the volumetric efficiency. *Volumetric efficiency* is really an incorrect description of what is being measured. But, the term has been in use for so many years that there's no real reason to try to change the usage to the correct term, *mass efficiency*.

Volumetric efficiency is the ratio of the *actual* mass (weight) of air taken into the engine—compared to the mass which the engine displacement would *theoretically* take in *if there were no losses.* This ratio is expressed as a percentage. It is quite low at idle and low speeds because the "pump" or engine is being throttled.

$$V.\ E. = \frac{Actual\ mass\ of\ air\ taken\ in}{Theoretical\ mass\ of\ air\ which\ could\ be\ taken\ in}$$

Volumetric efficiency reaches a maximum at a speed close to that where maximum torque at wide-open throttle occurs, then falls off as engine speed is increased to peak RPM.

The volumetric-efficiency curve closely follows the torque curve.

EFFECT OF V.E.

After figuring air-flow requirement for an engine with 100% volumetric efficiency, then you get realistic about your engine and reduce the air-flow number according to the volumetric efficiency you expect your engine to have.

Select a carburetor rated to flow the amount of air your engine actually needs, taking volumetric efficiency into account.

An ordinary low-performance engine has a V.E. of about 75% at maximum speed; about 80% at maximum torque. A high-performance engine has a V.E. of about 80% at maximum speed; about 85% at maximum torque. An all-out racing engine has a V.E. of about 90% at maximum speed; about 95% at maximum torque. A highly tuned intake and exhaust system with efficient cylinder-head porting and a camshaft ground to take full advantage of the engine's other equipment can provide such complete cylinder filling that a V.E. of 100%—or slightly higher—is obtained *at the speed for which the system is tuned.*

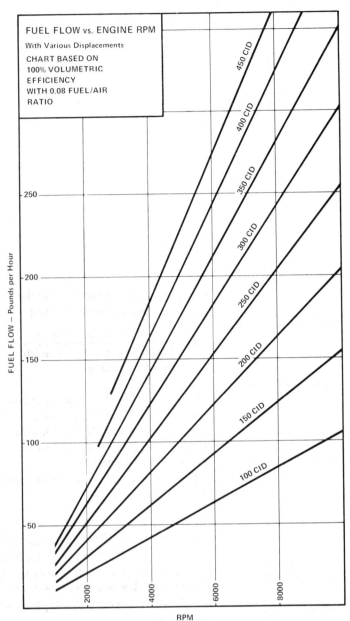

If you use this graph to estimate fuel flow, multiply the indicated fuel-flow rate by the volumetric efficiency you expect from your engine. This graph is based on a full-power fuel/air ratio of 0.08 which is suitable for nearly all engines.

NOTE: 1728 cubic inches per cubic foot. Displacements in cubic centimeters can be divided by 16.4 to convert them to cubic inches for use in these formulas.

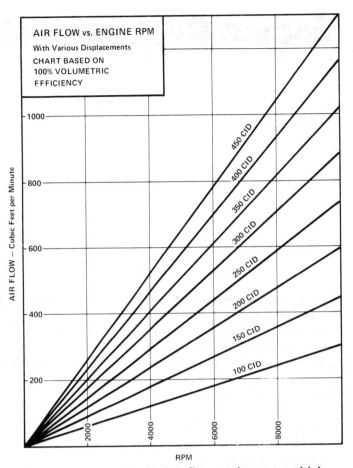

If you use this graph to find air-flow requirement, multiply the indicated air flow by the volumetric efficiency you expect from your engine. Use a carburetor with an air-flow rating equal to or slightly smaller than the air-flow requirement for your engine. The text provides further details.

STANDARD TEMPERATURE AND PRESSURE—High air temperature and high altitude reduce engine performance because both reduce air density. When air is less dense, an engine gets less usable air into the cylinders, so makes less power.

Automotive test engineers eliminate air temperature and pressure as a variable by correcting all readings such as air flow and engine power to what they would be at a standard temperature and pressure.

Standard temperature is 59° Fahrenheit. Standard pressure is sea-level air pressure which many people know is 14.7 pounds per square inch (psi). Carburetor engineers measure air pressure in inches of mercury (in. Hg) instead of psi. Standard air pressure is 29.92 in. Hg. One inch of mercury is approximately 0.5 psi.

Let's go back to our example of the 350 CID engine we calculated as flowing 811 CFM of air (at standard temperature and pressure) at 100% volumetric efficiency. If this is a high-performance engine with a maximum 85% V.E., then the flow becomes 811 CFM x 0.85 = 689 CFM (at standard temperature and pressure).

AIR MASS AND DENSITY

Because the mass of air taken in is directly related to the density of the air, volumetric efficiency can be expressed as a ratio of the density (γ) achieved in the cylinder versus the inlet density, or

$$\eta = \gamma_{cyl} / \gamma_i$$

Ideal mass flow for a 4-cycle engine is calculated by multiplying

$$\frac{RPM}{2} \times CID \times \gamma_i$$

where γ_i is the inlet air density.

Air density varies directly with pressure. The lower the pressure, the less dense the air. At altitudes above sea level, pressure drops, reducing power because the density is reduced.

Tables which relate air density to pressure (corrected barometer) and temperature are available. Or, you can use this formula:

$$\gamma = \frac{1.326P}{t + 459.6}$$

where

γ = density in lbs/cubic foot

P = absolute pressure in inches of mercury (Hg) read directly off of barometer (corrected)

t = temperature in degrees Fahrenheit at induction-system inlet

Actual mass flow into an engine can be measured as the engine is running by using a laminar-flow unit or other gas-measuring device. A calibrated orifice or a pitot tube can also be used.

Actual mass flow is usually lower than ideal in a naturally-aspirated engine because air becomes less dense as it is heated in the intake manifold. Absolute pressure (and density) also drop as the mixture travels the path from the carburetor inlet into the combustion chamber. This further reduces the mass of the charge reaching the cylinder.

The greater the pressure drop across or through the carburetor, the lower the density can be inside of the manifold and in the combustion chamber. If the carburetor is too small, pressure drop at wide-open throttle will be greater than the desired 1 inch Hg (for high performance) and power will be reduced because the mixture will not be as dense as it needs to be for full power.

Carburetor flow capacity is one way to state equivalent size. It is the quantity of air flow through the carburetor (at standard temperature and pressure) at a given pressure drop: Usually 1.5 inches of mercury for four-barrel carburetors and 3.0 inches of mercury for one- and two-barrel carburetors. The higher the flow rating—the bigger the carburetor. Or, the bigger the carburetor—the lower the pressure drop across it at any given air flow.

To keep volumetric efficiency as high as possible, we would like to keep the size of the carburetor up, hence the pressure drop down. *The limiting factor is at the other end of the flow curve.* Will the carburetor be able to meter fuel correctly at low air flows? Will it work o.k. at the most-often-used speed range for the engine?

FUEL REQUIREMENTS

Fuel requirements relate to the air-flow requirement because fuel is consumed in proportion to the air taken in by the engine. Fuel flow is stated in pounds per hour (lbs./hr.) and sometimes in lbs./HP hr., termed *specific fuel* consumption because it states how much fuel is used for *each* horsepower in one hour.

The relationship between the amount of fuel and the amount of air which flow together into an engine is called the fuel-air ratio. This is pounds of fuel divided by pounds of air. An engine uses a lot more air than fuel, so fuel-air ratios are always small numbers such as 0.08.

Some people find this easier to understand if the ratio (0.08) is turned upside down to become an air-fuel ratio. This gives a number like 12.5, meaning the engine uses 12.5 pounds of air for each pound of fuel. Obviously calculations can be made either way because these ratios are just two different ways to say the same thing.

First, let's look at the wide-open-throttle full-power fuel requirement for the same engine which needed 811 CFM

Carburetor Flow Rating in CFM — This is a better way to state carburetor capacity than using the older method of comparing venturi sizes. This is because venturi size does not accurately represent the actual flow capacity of the carburetor, especially where there are one or more booster venturis reducing the effective opening and increasing the restriction or pressure drop across the carburetor.

EQUIVALENT RATIO TABLE

A/F (Air/Fuel)	F/A (Fuel/Air)	A/F (Air/Fuel)	F/A (Fuel/Air)
22:1	0.0455	13:1	0.0769
21:1	0.0476	12:1	0.0833
20:1	0.0500	11:1	0.0909
19:1	0.0526	10:1	0.1000
18:1	0.0556	9:1	0.1111
17:1	0.0588	8:1	0.1250
16:1	0.0625	7:1	0.1429
15:1	0.0667	6:1	0.1667
14:1	0.0714	5:1	0.2000

Holley engineering uses several air-box rooms. Tall cylinders supply air to air boxes hidden behind the cylinders. Air box at right foreground uses orifice entry to air box (created when cover is lowered over carburetor). Vacuum is provided by giant vacuum pumps. Air-box instrumentation includes mercury and water manometers for measuring pressure drop and for checking vacuums at various parts of the carburetor. Air flow directly read out in CFM from timed meter is converted to pounds per hour from a chart and used with the fuel-flow measurement (in pounds per hour) to calculate F/A ratio at various flow rates and throttle openings. Fuel pressure is read directly from a gage and a burette provides an accurate indication of accelerator-pump-delivery capacity. Console at left is one of many computerized carburetor test stands used by Holley for automatically testing all production carburetors.

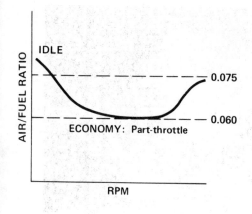

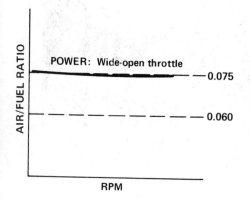

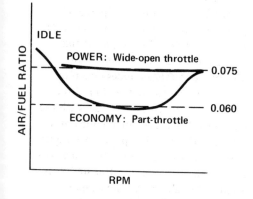

of air at 8,000 RPM at 100% volumetric efficiency. To get the air flow into lbs./hr.:

CFM x 4.38 = Air Flow lbs./hr.

Taking this a step further, multiply by the fuel/air ratio, which we will assume to be a typical full-power ratio of 0.077 lbs. fuel/lbs. air (air/fuel ratio of 13:1).

CFM x 4.38 x F/A = Fuel Flow lbs./hr.

811 x 4.38 x 0.077 = 273 lbs./hr. @ 8,000 RPM

This is the maximum fuel the engine could consume with 811 CFM air flow, but this fuel flow will hardly ever be reached because—like air flow—fuel flow must be reduced by volumetric efficiency:

Assuming η = .85

CFM x 4.38 x F/A x η = Fuel Flow lbs./hr.

811 x 4.38 x .077 x 0.85 = 232 lbs./hr. @ 8,000 RPM

And fuel flow is less when the engine is turning fewer RPM. For instance, 232 lbs./hr. @ 8,000 RPM drops to 1/2 or 116 lbs./hr. at 4,000 RPM and 58 lbs./hr. at 2,000 RPM. . .assuming wide-open throttle in each instance, and the same fuel-air ratio.

The accompanying chart on page 7 shows fuel flow for various engine sizes over typical RPM ranges. If you don't happen to have this book with you when you need to estimate fuel requirements for fuel pump and fuel line selection, just remember: Wide-open-throttle full power typically requires 0.5 lb. fuel per HP every hour. Thus, a 300 HP engine needs 300 x 0.5 = 150 lbs./hr., or 25 gallons per hour because gasoline weighs 6 pounds per gallon.

The preceding discussion is related to the maximum amount of fuel the engine could be expected to consume under full-power, wide-open-throttle conditions. This knowledge will help you in selecting fuel-line sizes, fuel-pump capacity and the required tank capacity to get your vehicle where you want it to go within the limitations of fuel stops. A lot of high-performance enthusiasts blame the carburetor for leaning out when the real problem is one of inadequate fuel supply. If you're in a hurry to find out more about this, rush on back to the fuel-supply system chapter.

Now let's look at fuel requirements for other conditions.

Stoichiometric or Chemically Correct Mixture—You've heard of *ideal* situations where everything is absolutely perfect? Well, this is the *ideal* situation in an engine whereby all of the fuel is perfectly mixed and completely burned. The residue or exhaust is carbon dioxide, water vapor and nitrogen.

Fuel + Air ⟶ CO_2 + H_2O + N_2

This would be a fuel/air (F/A) ratio of approximately 0.068, or an air/fuel (A/F) ratio of 14.7:1.

The actual ratio at which this occurs with an ideal set of conditions varies with the fuel's molecular structure. Gasolines will vary somewhat in structure, but not significantly. Fuels other than gasoline require different ratios for the ideal or stoichiometric condition. Alcohol, for instance, has a lower heat content (calorific value) than gasoline and requires a 0.14 F/A ratio for its ideal burning condition. This is so much more fuel volume than required for gasoline that most carburetors cannot be used with alcohol, even when highly modified. Passages in the carburetor are actually too small to allow correct fuel flow and metering.

Maximum Power—Maximum power requires excess fuel to make sure all of the oxygen in the air is consumed. The reason for excess fuel: Mixture distribution to the various cylinders and fuel-air mixing are seldom perfect. When all of the air enters into the combustion process, more heat is generated and heat means pressure the engine can convert to work.

The reaction looks like this:

Fuel + Air ⟶ CO_2 + H_2O + CO + HC + N_2

where:

CO_2	=	carbon dioxide
H_2O	=	water
CO	=	carbon monoxide
HC	=	unburned hydrocarbons (gasoline)
N_2	=	nitrogen

The fuel excess usually amounts to 10 to 15%, giving fuel/air ratios 0.075 to 0.080 (13.3 to 12.5:1 air/fuel ratios). Sometimes, an excess of fuel beyond that which produces maximum power is used for internal cooling of the engine. From a pollution standpoint, unburned hydrocarbons and carbon monoxide are undesirable by-products.

Maximum Economy—Maximum economy requires excess air to ensure that all of the fuel is consumed. Typical F/A ratio is 0.06. The reaction looks like this:

$$Fuel + Air \rightarrow CO_2 + H_2O + N_2$$
$$+ \text{ small amounts of}$$
$$CO + HC + O_2$$

Under heavy loads, any mixture leaner than stoichiometric burns with sufficient heat to cause any free oxygen to be combined with the remaining nitrogen to produce oxides of nitrogen (NO_x). This is one of the undersirable emission products.

At high engine speeds and low loads, mixtures approaching 0.055 F/A (18:1 A/F) are sometimes approached when seeking peak economy. However, this is on the borderline where fuel/air begins to become unstable in its ability to burn in normal engines.

Idle—At idle, the problem of exhaust dilu-iton appears. Dilution of the intake charge is caused by the high manifold vacuum and by opening the intake valve before the piston reaches top center. When the intake valve opens, some exhaust gases are forced into the intake manifold by the pressure difference, assisted by the still-upward-rising piston. These dilute and effectively lean the charge. Also, some of the exhaust gas stays in the cylinder clearance volume. When the piston descends on the intake stroke, the initial charge consists largely of exhaust gas, with a small proportion of fresh fuel/air mixture. As the piston completes its intake stroke, the percentage of fresh fuel/air mixture is usually somewhat higher and this portion of the inlet charge burns well enough to supply power for idling the engine.

In the process of dilution, some fuel molecules combine (or line up, as the chemists say) with the exhaust molecules and some of the fuel molecules line up with the oxygen in the air. To make sure that the mixture is combustible, the mixture has to be made 10 to 20% richer than stoichiometric to offset that part of the fuel which combines with the exhaust gas. The richer mixture also helps to offset distribution problems. It also creates additional emission problems because rich idle mixtures generate large amounts of carbon monoxide (CO).

Thermocouples inserted in headers (arrows) measure exhaust temperatures for each cylinder to assess distribution characteristics of manifold/carburetor combination.

Cold Starting—Starting a cold engine requires the richest mixtures of all because slow cranking speeds provide air velocity too low for much fuel vaporization to occur. And, because both the fuel and the manifold are cold, fuel vaporizes poorly. Vaporization is necessary for combustion, so a lot of excess fuel is required for starting. Typical fuel/air ratios are 1.0 to 2.0. The fuel/air ratio being supplied is not that which is getting to the cylinders. Distribution problems, depositing of liquid fuel on the walls of the manifold and on the ports in the cylinder heads all rob fuel from the mixture so the mixture actually reaching the cylinders in a vaporized form is probably in the 0.06–0.10 range.

Once the engine fires, RPM goes up, velocity through the carburetor increases so there is better vaporization and fuel which has been deposited on the manifold walls in a liquid state vaporizes. As this happens, the mixture is gradually leaned out to 0.12–0.14 F/A, still 80 to 120% richer than nominal. When the engine warms up, normal operating mixtures can be used.

UNIFORM DISTRIBUTION

Now that we have looked at the engine's fuel and air requirements, let's consider another important engine requirement: Uniform (or equal) fuel-air mixture distribution to the various cylinders. Uniform distribution helps every phase of engine operation. Non-uniform (unequal) distribution causes a loss of power and efficiency, and may increase emissions.

Uniform distribution relies on a number of factors: Good atomization of the fuel by the carburetor, use of fuel with the correct volatility for seasonal temperature variations, utilizing appropriate techniques to ensure vaporization of the fuel, and correct manifold design. This latter subject is covered in detail in a separate chapter later in the book. We will discuss the other items after we have established the importance of uniform distribution.

Uniform Distribution at Idle—As discussed in the preceding fuel-requirements section, idle F/A is typically 10 to 20% richer than ideal to offset dilution by exhaust gas. Idle mixtures with less than 1% CO

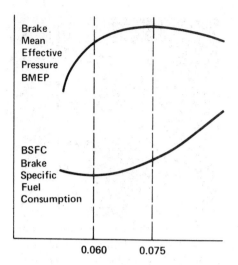

Brake mean effective pressure (BMEP) of engine operating with perfect distribution, showing relationship of F/A ratio supplied by carburetor. For maximum power, 0.075 to 0.080 F/A gives lowest possible fuel consumption for that output. For maximum economy, 0.060 F/A gives lowest possible fuel consumption consistent with a lower, best-economy power output.

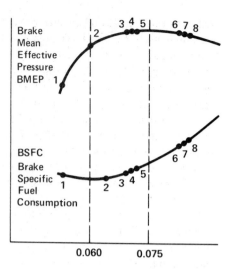

Same curve as at left with numbers representing cylinders receiving various F/A ratios at maximum-power output with non-perfect distribution. All cylinders are sharing the same carburetor and manifold. 6,7,8 with too-rich F/A produce less power and consume excess fuel. 3,4,5 receiving correct F/A for best power consume least fuel for that output. 1,2 have lean F/A ratio. A richer mixture would have to be supplied to ALL cylinders to bring 1 & 2 to the flat part of the curve so as to avoid detonation. The net result: Only cylinders 1 & 2 would operate at peak power with minimum fuel consumption for that output. 3,4,5,6,7, 8 would produce less power than if the engine had perfect distribution—and consume more fuel. If the curve is assumed to represent distribution during economical part-throttle operation, cylinder 1 receiving a lean F/A will operate at high fuel consumption due to misfiring. The other cylinders will operate at ratios above the correct 0.060 F/A, wasting fuel.

(a common emission setting) require that all cylinders receive approximately the same idle mixture.

If distribution is not good, getting a reasonably smooth idle may require idle-mixture settings which are too rich. This can give CO emissions too high to meet emission regulations.

Uniform Distribution for Economy—Maximum economy, as previously discussed, requires a F/A ratio of approximately 0.06. If one cylinder receives a lean charge due to a mixture-distribution problem, this can increase the amount of fuel required for a given power output. To avoid misfiring in the lean cylinder, the other cylinders will have to be operated at mixtures richer than the desired 0.06 F/A to bring the lean one up to the 0.06 figure.

All cylinders should be operating at the same F/A because lean cylinders produce more oxides of nitrogen (NO_x) if they are still firing. And, if they are so lean that misfiring is occurring, unburned hydrocarbons are emitted. Rich-running cylinders will provide excess CO and unburned hydrocarbons. In either case, undesirable effects include decreased economy and increased emissions.

Uniform Distribution for Power—F/A ratios of approximately 0.075 are needed to produce maximum power. If one cylinder runs lean, an excessively rich mixture must be supplied to the other cylinders to bring the lean cylinder up to the correct ratio so it will not run into detonation. Rich cylinders will waste a lot of fuel and fuel consumption will be increased to maintain a specified power level. Running some cylinders rich reduces their power output because all cylinders must have the same F/A to get the best possible power out of the engine.

Unequal distribution not only affects fuel consumption and power—it also means ignition timing can only be a rough approximation or compromise of what could be used if all cylinders received equal F/A mixtures. Spark advance which can be used is directly related to the F/A available in the cylinder. Advance is typically limited to that which will not produce knock in the leanest cylinder. This limits the power which can be developed from those cylinders which have been richened to compensate for unequal distribution.

Checking for Correct Distribution—Automobile engine designers—and designers of carburetion systems, too—use a number of methods to check whether the various cylinders in an engine are receiving a uniform mixture. It should be noted that all of these methods require the use of a chassis or engine dynamometer. Generally speaking, these methods are not often used by individual tuners because of the cost and complexity of the equipment. In many instances combinations of methods are used. Distribution is checked by:

1. Chemically analyzing exhaust-gas (combustion products) samples from the individual cylinders at various operating modes: Idle, cruise, acceleration, maximum power and maximum economy. There is a definite relationship between F/A and the chemical components in the exhaust gas. By using the very accurate analytical equipment developed for emission studies, exact determinations of F/A ratio are obtained to check distribution. The availability of such equipment has made chemical analysis the method preferred by automotive engineers for distribution studies.

2. Measuring exhaust-gas temperature (EGT) with thermocouples inserted into the exhaust manifold or header at each cylinder's exhaust port. The designer looks for EGT peaks as various main-jet sizes are tried. If all cylinders were brought to the same temperature, one (or more) cylinder might be 200°F below its peak (and therefore below its peak output) because all cylinders do not produce the same EGT due to differences in cooling, valves, porting and ring sealing. Perfect distribution would place all cylinders at their peak EGT's with the same jet—*an ideal situation which is rarely achieved.*

3. Studying the specific fuel consumption to determine how closely the F/A delivered by a particular carburetor and manifold match ideal F/A ratios at various operating conditions. Any great variance is cause for suspecting a distribution problem.

4. Observing combustion temperatures at various operating conditions.

FACTORS AFFECTING DISTRIBUTION
Atomization & Vaporization—Before gasoline can be burned, it must be vaporized. Vaporization changes the liquid to a gas

state and this change only occurs when the liquid absorbs enough heat to boil. For example, a tea kettle changes water to water vapor (steam) by transferring heat into the water until the boiling temperature is reached. At this point, additional heat must be added to change the water into steam and the steam enters the atmosphere as water vapor (a gas, really). This extra heat is called the *heat of vaporization.* Pressure controls the boiling point of any liquid. In the case of water, 212°F (100°C) is the sea-level boiling point. At higher altitudes, due to lowered atmospheric pressure, water boils at lower temperatures. Remember this relationship because it is important in understanding what happens in the carburetor and manifold.

In most passenger-car engines, heat is applied to the intake manifold to raise the temperature of the incoming mixture.

The higher the temperature the better the vaporization. There will be some sacrifice in top-end power when the mixture is heated sufficiently to ensure vaporizing most of the fuel. This is due to the reduction in charge density which occurs when the mixture is heated to this point. In most passenger-car engines, the loss of power is more than offset by the smoother running gained at part throttle. No manifold heat is used in a racing engine because a cold, dense mixture produces more power.

Not only is it necessary to vaporize fuel before it can be expected to burn—it is essential to vaporize it to aid in distribution. It is much easier to distribute vaporized fuel in a fuel/air mixture than it is to distribute liquid fuel. It should be noted that some liquid fuel is nearly always present on the manifold surfaces. The only time the manifold surfaces will be "dry" is during high-manifold-vacuum conditions which promote vaporization.

Gasoline in the carburetor is discharged into the air stream as a spray, which is atomized (torn or sheared into fine droplets) into a mist. At this point it should become vaporized. Pressure in the intake manifold will be much lower than atmospheric (expect at wide-open throttle) and this considerably lowers the boiling point of the gasoline. At the reduced pressure, some fuel particles are vaporized as they absorb heat from the surrounding air.

Pinto engine plumbed for exhaust sampling to check distribution and emission characteristics. Engine is on one of five engine dynamometers Holley uses for carburetor development. Holley 5200 is the original equipment carburetor on this engine.

And, some fuel particles are vaporized when they come in contact or close proximity to the hot spot on the manifold floor.

Imperfect vaporization may occur if the mixture velocity is too low, if the manifold or incoming air is cold, if manifold vacuum is low (higher pressure), and if the fuel volatility is too low for the ambient temperature. Vaporization is also affected by manifold design and the carburetor size (flow capacity). With the same engine speed, a large carburetor has less velocity through its venturi (air passage), hence less pressure drop and a greater tendency for the fuel to come out of the discharge nozzle in liquid blobs or large drops which are not easily vaporized.

Manifold design also enters into the vaporization picture because the size of the passages affects the mixture velocity and heating. If the mixture travels slowly, some liquid particles of fuel may deposit onto the manifold walls before they have a chance to vaporize. The hot-spot size and location and the surface area inside the manifold dramatically influence vaporization.

When vaporization is poor, an excess of liquid fuel gets into the cylinders.

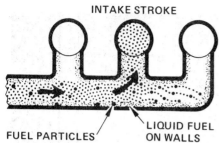

INTAKE STROKE

FUEL PARTICLES
LIQUID FUEL ON WALLS

When the mixture is only partially vaporized, liquid particles tending to cling to the manifold walls—or avoiding sharp turns into a cylinder may cause some cylinders to run lean and some rich. In this example, the center cylinder will tend to be leaned and the end cylinder will tend to receive a rich mixture. Where possible, it is best to divide the mixture before a change in direction occurs. Liquid particles are relatively heavier than the rest of the mixture and tend to continue in one direction. A fully vaporized mixture promotes good distribution in every instance.

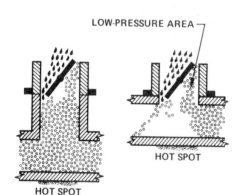

LOW-PRESSURE AREA

HOT SPOT

HOT SPOT

LOWER LEVEL CROSS-H MANIFOLD

UPPER LEVEL CROSS-H MANIFOLD

Cross section through V-8 engine manifold of Cross-H or two-level type shows how riser height can affect distribution. Throttle shown in worst position causing greatest effect on distribution.

Because it does not completely burn due to the lack of time available for evaporation and burning, it is expelled as unburned hydrocarbons. Some excess fuel washes oil off of the cylinder walls to cause rapid wear. And a portion of the liquid fuel drains past the rings into the crankcase where it dilutes the oil.

Exhaust-heated hot spots are typically small areas just under the area which is fed by the carburetor. The ends of the manifold are not usually heated. The size is kept as small as possible, consistent with the needs for flexible operating and smooth running. By keeping the spot fairly small, the manifold automatically cools off as RPM is increased. The large amount of fuel being vaporized at high speed extracts heat from the manifold—often making it so cold that water condenses on its exterior surfaces. Although most passenger-car manifolds heat the mixture with an exhaust-heated spot, some manifolds are water-heated by engine coolant. Cars equipped with emission controls often heat the incoming air by passing the air over the exhaust manifold on its way to the air-cleaner inlet.

Excepting racing intake systems, intake manifolds are compromise devices. Their shape, cross-sectional areas, and heating arrangements accomplish the necessary compromises between good mixture distribution and volumetric efficiency over the range of speeds at which the engine will be used. If only maximum or near-maximum RPM is being used, high mixture velocity through the manifold will help to ensure good distribution and will help to vaporize the fuel—or at least hold the smaller particles of fuel in suspension in the mixture. At slower speeds, the use of manifold heat becomes essential to ensure that the fuel is vaporized. If heat is not used, the engine will become rough running at slower speeds and distribution problems will be worsened.

Fuel Composition—The more volatile the fuel, the better it will vaporize. All fuels are blends of hydrocarbon compounds and additives. The blending is typically accomplished to match the fuel to the ambient-temperature and altitude conditions. Thus, fuel supplied in the summertime has a higher boiling point than that which is available in winter. Fuel volatility is rated by a number known as *Reid Vapor Pressure.* In Detroit, for example, the

Reid number varies from 8.5 in the summer to 15 in winter. Fuel blending is always a compromise made on the basis of estimates of what the temperature will be when the gasoline is used. A sudden temperature change—such as a warm day in winter—usually causes a rash of vapor-lock and hot-starting problems.

Carburetor Placement—Location of the carburetor on the manifold in relation to the internal passages can drastically affect distribution. If the geometric layout of the manifold places the carburetor closer to one or more cylinders, this can create problems. Considerations of carburetor location become especially important with multiple carburetors.

It's a Combination—The engine designer checks power, economy and mixture distribution with the carburetor, air cleaner and manifold installed. Sometimes the physical positioning of the air cleaner or a connecting elbow atop the carburetor is so critical that turning it to a different position can create distribution problems.

Throttle-Plate Angle—A directional effect is given to the fuel when the throttle is partially open. This effect becomes especially apparent when the manifold has little or no riser between the carburetor and the manifold passages. A riser is the vertical passage which conducts mixture from the carburetor into the intake manifold.

Throttle-plate angle affects distribution in the upper level of a cross-H (or two-level) manifold (for a V-8 engine) more than it does the lower level because the longer riser into the lower level has a straightening effect on the mixture. This points up one of the advantages of high-riser designs which allow better straightening of the mixture flow with less directional effects at all throttle openings. High-riser designs typically offer better cylinder-to-cylinder distribution of the mixture. While height is usually limited by hood clearance, Holley tests have shown definite improvements in distribution at part throttle with risers 1-1/2 to 2 inches high prior to opening into the manifold branches.

Mixture Speed & Turbulence—Mixture velocities and turbulence within the manifold definitely affect vaporization and hence distribution. More details are in the manifold chapter. Turbulence in the combustion chamber helps to prevent stratification of the fuel and promotes rapid flame travel.

Time—The volatility of the fuel and the heat available to assist in vaporization are especially important when you consider the tiny amount of time available for vaporization of the atomized fuel supplied by the carburetor. Unlike water in a tea kettle which can be left on the stove until it boils—fuel must be changed to the vapor state in about 0.003 second in a 12-inch-long manifold passage at a mixture velocity of 300 feet per second.

DISTRIBUTION SUMMARY

To summarize the factors which can aid distribution:

1. Vaporize as much fuel as possible in the manifold so a minimum of liquid fuel gets into the cylinders.

2. Use fuels with the correct volatility for the ambient-temperature and altitude conditions.

3. Ensure high velocities in the manifold by using the smallest passages consistent with the desired volumetric efficiency.

4. Provide good atomization in the carburetor through careful selection of venturi size (carburetor flow capacity).

5. Avoid manifold construction which causes fuel to separate out of the mixture due to sharp turns and severe changes in cross-sectional area.

6. Provide sufficient turbulence in the manifold to ensure fuel and air are kept well mixed as they travel to the cylinder.

7. Provide manifold heating and/or heating of the incoming air to ensure that fuel is well vaporized.

Obviously, not all of these factors can be controlled by an individual. But, they bring out a few important points such as avoiding carburetors which are too large for the engine. Use fuel which is correct for the season; don't try to race with fuel you bought in a different season or another locality. And, for engines being operated on the street, use a heated manifold with the smallest passages consistent with the desired performance.

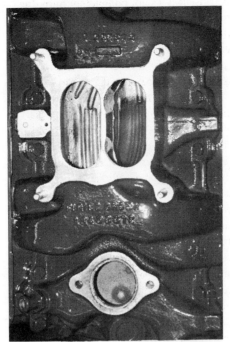 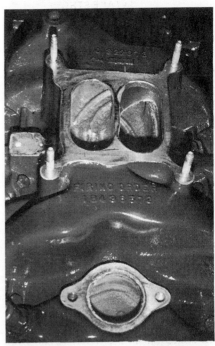

Two views of Chevrolet high-rise manifold which is supplied as a stock item on the Z-28 and LT-1 engines. The two distinct levels below the flange identify this as a cross-H design. Ribs in manifold floor are distribution devices to direct mixture and to aid in controlling liquid fuel when vaporization is not perfect. Details on modifying the center divider and using a spacer for increasing high-RPM power are covered in H. P. Books' *How to Hotrod Small-Block Chevys.* Manifold design is covered in an entire chapter starting on page 102.

How your carburetor works

Dual-lung float assembly of Model 5200 has bumper spring (arrow). Spring-loaded inlet valve is hooked to lever for positive opening of valve when float drops (outline arrow). Brass wire-mesh filter catches dirt which could cause inlet-valve malfunctioning. Large tube is fuel inlet, smaller is fuel return for purging vapors from fuel line. There should be 0.010 to 0.025-inch clearance between tang of float and bumper spring when bowl cover is inverted and float is resting against spring-loaded needle as shown here.

Understanding how your carburetor works is the key to get top performance from your engine/carburetor combination. And, it's also the quick way to get more performance with less work. You will know what is happening and be able to do your tuning in a meaningful way without wasting time. Cut-and-try changes can wreck your carburetor and/or leave your combination in a worse state

of tune than it was when you started. Also, once you know what happens inside of a carburetor, you will be able to choose the right carburetor for your engine without having to rely on "experts" or rumors about the current "hot setup."

There's really nothing "tricky" about how your carburetor works. And, no "black magic" makes one work differently from another. Whether you have a one-,

two- or a four-barrel carburetor, all operate essentially the same as all others. There are minor differences related to the way the same kinds of systems were built into a particular carburetor. Carburetors are really very simple and similar devices, as you will see. Just as all four-cycle engines—from one-lung lawnmowers to Chrysler hemis—operate with the same principles of intake, compression, power and exhaust, all carburetors operate similarly when you compare them. A four-barrel double-pumper may look impressive and complex, but it works exactly like a one-barrel—and its workings are certainly no harder to understand. Learning how your carburetor works is so simple you won't believe it until you finish reading this chapter.

Regardless of how many venturis (barrels) your carburetor has, let's keep things simple by starting out with a one-barrel carburetor, just as we learned about engines by examining what goes on in a single cylinder. After you have become completely familiar with the one barrel—then we'll proceed to the two- and four-barrel units. Don't jump ahead because you have to learn what goes on in one barrel before you try to comprehend more complex carburetors which are literally several one-barrel carburetors built into a single unit.

The carburetor is constructed to do several jobs:

1. It controls engine input and therefore controls the power output.

2. It mixes fuel and air in the correct proportions for engine operation.

3. It atomizes and vaporizes the fuel/air mixture to put it in a homogeneous state for combustion. Let's look at the basic systems in the carburetor one at a time and see what parts do each job. These show the relationships involved and how they all work together. After reading this material you will be well on the road to a complete understanding of how your carburetor—or *any* carburetor—works.

Inlet System

DESCRIPTION OF OPERATION

The inlet system consists of three major items:

1. Fuel bowl
2. Float
3. Inlet valve (needle and seat)

Fuel for the basic metering systems of the carburetor is stored in the fuel bowl. The fuel-inlet system must maintain the specified fuel level because the basic fuel-metering systems are calibrated to deliver the correct mixture only when the fuel is at this level. Correct fuel level also greatly affects fuel handling—the carburetor's ability to withstand manuevering: Quick accelerations, turns, and stops.

The amount of fuel entering the bowl through the fuel-inlet valve is determined by the space (flow area) between the movable needle and its seat and by fuel-pump pressure. Movement of the needle in relation to its seat is controlled by the float which rises and falls with fuel level. As fuel level drops, the float drops, opening the needle valve to allow fuel to enter the bowl. When the engine is running with constant load, the float moves the needle to a position where it restricts the flow of fuel, admitting only enough fuel to replace that being used. Any slight change in the fuel level causes a corresponding movement of the float, opening or closing the fuel-inlet valve to restore or maintain the correct fuel level.

DESIGN FEATURES/Fuel Inlet System

Fuel Bowl—The fuel bowl or float chamber is a reservoir which supplies all fuel to the carburetor. The height of fuel in the bowl is controlled by the float and inlet valve as described above. The pressure of fuel in the bowl is approximately the same as outside air because an air passage, or vent, connects the float bowl to the air inlet passage for the carburetor.

Venting the fuel bowl to outside air means that the fuel is no longer pressurized by the fuel pump. It is pressurized up to the float valve, but once it is in the fuel bowl, it is at *vented pressure*.

The bowl also acts as a vapor separator. Vapors which may have been entrapped in the fuel as it was pumped from the

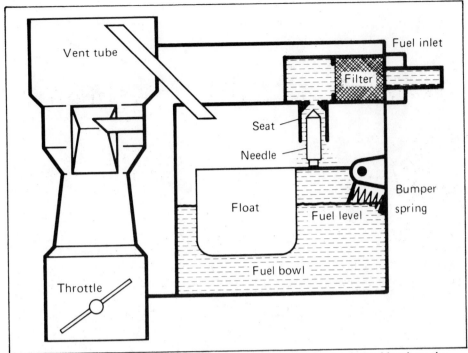

Schematic of the inlet system. Here the float has not reached the desired level, so the inlet valve is off its seat and admitting fuel to the bowl. Fuel flow will shut off when the float rises to close the inlet valve.

tank escape through this vent so there is no pressure build-up in the bowl.

By connecting the vent to the inlet air horn, the fuel bowl is vented to clean air obtained through the air cleaner. And, because the vent "sees" the same pressure as the carburetor inlet, a dirty air cleaner does not change the F/A ratio. Dirty air cleaners restrict the air flow, creating a lower absolute pressure to the carburetor, which causes a power loss. If the fuel bowl is not vented to the inlet air horn, a dirty air cleaner will enrich the mixture.

The preferred location for a vent is near the center of the fuel bowl—high enough so fuel will not slosh into the air horn during hard stops or manuevering. Some 1970 and later cars have an additional vent connecting the float bowl to a charcoal canister when the engine is turned off. This part of the emission-control system collects escaping fuel vapors generated when heat from the engine block warms the carburetor, causing gasoline in the fuel bowl to "boil" off. The same canister may also collect fuel-tank vapors. Vapors are sucked back into the intake manifold from the canister when the engine is running. On pre-emission-control models there was often a mechanical vent valve

on the fuel bowl. At curb idle or when the engine was stopped, this external vent released fuel vapors into the engine compartment.

The fuel bowl may be an integral part of the main body casting, or it can be a separate casting attached to the carburetor body with screws. Holley uses both designs. Both fuel-bowl types have replaceable screw-in inlet valves.

Bowl capacity is important. It must contain sufficient fuel to allow good response when accelerating from stop after a hot engine has been idling or stopped (called *hot soak*). At such times the fuel pump may deliver spurts of liquid fuel and pockets of fuel vapor intermittently. Thus the bowl must be large enough to supply fuel to the metering systems until the pump is purged of vapor and able to deliver liquid fuel. Holley carburetors typically have large-capacity bowls. Sometimes bowl capacity is reduced with bowl inserts or *stuffers* to allow the vehicle to pass evaporative-emission requirements.

Float—A float on a hinged lever operates the inlet valve so fuel enters the bowl when the fuel level is below the desired reference height—and shuts off fuel when

Three types of Holley removable fuel bowls: Left is center-hung or "race" type, center is side-hung—both can have the fuel level adjusted without removing the bowl. At right is now-obsolete "nose" bowl. It was once widely used. Center bowl has the lock screw and adjusting nut installed to allow positioning inlet valve to set float and fuel level.

Side-hung float in secondary fuel bowl shows float 1, adjustable needle/seat assembly 2, adjusting nut 3, lock screw 4, plastic baffle 5 to direct inlet fuel and contain froth, float bumper spring 6, sight plug 7 for fuel-level checking and hole 8 for transfer tube which brings fuel to secondary bowl from inlet on primary bowl. When correctly set up, half-round float is adequate for all but the most severe cornering loads such as may be generated in autocrossing, road racing or on a flat asphalt track.

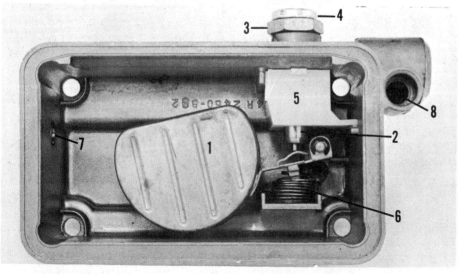

the desired height is reached. A float may have one or two buoyant elements, sometimes referred to as "lungs" because their shape is sometimes similar to that of the human lung. Floats are made from brass stampings soldered together into an air-tight assembly; or from a closed-cellular material which is not affected by gasoline, alcohol or any of the other commonly used fuels or fuel additives. The float is designed with adequate buoyancy so fuel will be positively shut off by the inlet valve when the desired fuel level has been reached. The buoyant portion is usually mounted on a lever to multiply the buoyancy effects of the float itself and provide a surface for operating the inlet valve.

A float spring is sometimes added under the float to minimize vibration of the float. Float vibration, caused by engine or vehicle vibration or bouncing—or both—can cause wide fuel-level variations because a "jumping around" float allows the inlet valve to admit fuel when it is not needed. These are sometimes referred to as *float bumper springs.* Some carburetors may use a tiny spring inside of the inlet valve itself instead of—or in addition to—a spring under the float. The Holley Model 1901 "Buggy Spray" is an example of a carburetor with both damping methods. Such springs can be especially helpful in dirt-track, off-road and marine applications.

Float shape and mounting (pivot orientation) may be dictated by bowl size and the use of the carburetor. The float has to provide enough buoyancy to close the inlet valve and this sometimes requires making a very odd shape to get enough float volume to provide the required buoyancy. In the case of the Holley Model 1901 "Buggy Spray" carburetor, the float and fuel bowl are concentric with the main air passage to minimize the effects of operation on inclines, hard stops and severe turns.

The length of the float lever determines the mechanical advantage which the buoyancy of the float (and float spring, if used) can apply to close the inlet valve at a given fuel level.

Every carburetor has a published float setting. The float setting is the float location which closes the inlet valve when the required fuel level in the bowl is reached.

In some Holley carburetors, a threaded inlet-valve assembly can be adjusted to set fuel level without taking the carburetor apart. In this instance, a sight plug in the fuel bowl can be removed to observe the fuel level.

Some float settings are established by measuring float position—usually in relation to a gasket surface—while the carburetor is disassembled.

A fuel level for a particular carburetor is established by the designer and test engineer so the carburetor will operate without problems in fast starts and stops and in manuevers which would ordinarily be encountered in the particular vehicle for which the carburetor was made. And, the level is set so there will be no fuel spillage when a passenger car is parked or operated facing up, down, or sideways on a hill with 32% grade (18°).

Another specification for military vehicles requires correct operation when parked or operated at 60% (31°) up/down and/or a 40% (22°) side slope.

For best operation in high-speed cornering, bowls are equipped with center-pivoted floats with the pivot axis parallel with the axles of the car. For best resistance to the effects of acceleration and braking, side-hung floats with pivots perpendicular to the axles have a slight advantage.

Two- and four-barrel carbs have primary venturis which are opened first and secondary barrels which are opened to air flow later when higher engine power is required. Often primary and secondary venturis (throttle bores) are served by separate float bowls.

To offset the slosh-over tendencies of the fuel as it moves toward the carburetor throttle bores on hard stops, fuel levels in rear secondary bowls are typically set 1/16- to 1/8-inch lower than the forward or primary bowls. This lower fuel level makes the fuel rise higher before it can spill over through the main discharge nozzle or vents—making the carburetor more tolerant of hard stops.

Inlet Valve—The rounded end of the needle rests against the float lever arm, or is connected to it by a hook. The tapered end closes against an inlet-valve seat as the float rises.

As mentioned in the float section, some inlet needles are hollow and contain a tiny damper spring and pin to help cushion the needle valve so that it is protected from road shocks and vehicle vibrations.

EFFECTS OF INLET SYSTEM CHANGES

CHANGE	EFFECT
High float level	Raises fuel level in bowl. Speeds up main system "start up" because less depression is required at venturi to start fuel "pullover." Increases fuel consumption. May cause fuel spillage through discharge nozzles and/or vent into carburetor air inlet on quick stops or turns — causing engine to run erratically or stall. Increases tendency for the carburetor to "percolate" or to "boil over" — a condition in which fuel is pushed by rising vapor bubbles out of the discharge from the main well in a hot-soak situation. Also, when the vehicle is parked on a hill or side slope, high float level may cause spillage through the vent or main system at an inclination angle less than the design specification.
Low float level	Lowers fuel level in bowl. Delays main-system "start-up" because more depression is required at venturi to start fuel "pull-over." This delay may cause flat spots or "holes." May expose main jets in hard manuevering, causing "turn cutout." Can lessen maximum fuel flow capability.
Bumper spring too strong (too heavy)	Causes low fuel level. Inlet valve closes prematurely because of added closing force of closing spring.
Bumper spring too weak (too light)	Causes high fuel level. Inlet valve closes after correct fuel level has been reached because additional float displacement is required to compensate for weaker bumper spring force.
High fuel pressure	Will raise fuel level approximately 0.020 inch per psi fuel-pressure increase.
Low fuel pressure	Will lower fuel level.
Larger inlet-valve seat	Will raise fuel level.
Smaller inlet-valve seat	Will lower fuel level.

NOTE: Changing almost anything in the fuel-inlet system makes it necessary to reset the float to obtain the correct fuel level.

Inlet-valve needles are usually made of steel with a tapered seating surface tipped with Viton. Viton-tipped needles are extremely resistant to dirt and conform to the seat for good sealing with low closing forces. Plain-steel needles are recommended for use with other fuels such as alcohol and nitromethane.

Inlet valves are supplied with various seat openings. Each inlet valve assembly (if the needle is an integral part) and each inlet-valve seat is stamped with a number indicating seat opening in thousandths of an inch. A valve assembly marked 110S has a 0.110-inch-diameter seat opening and a Viton-tipped needle; a 110 is the same seat opening with a steel needle. Seat diameter is an indication of how much fuel flow occurs at a given fuel pressure. A smaller opening flows less fuel— a larger opening, more. The chart on page 20 shows fuel flow with a 0.110-inch inlet seat at various fuel pressures with the needle fully open (float dropped all the way down).

Seat size is selected to allow reasonably quick filling of the bowl to handle quick accelerations after standing parked with a hot engine and to give minimum restriction for high fuel demands such as occur at wide-open throttle at high RPM. Larger seats are also used to provide better purging of vapor from the fuel lines. A small needle seat provides the best control of hot fuel because vapor pressure in the fuel line acts against less area to force the needle off of its seat. A needle seat which is too large is a definite hindrance in off-road work and in other types of racing, too. A small needle seat aids in controlling the fuel. Any change in seat size requires resetting the float to the float height specification.

A fuel filter or screen may be included in the carburetor body, fuel-bowl or cover as part of the fuel-inlet system. The filtering device is placed between the fuel pump and the inlet valve to trap dirt which could cause inlet-valve-seating problems. Fuel-line fuel filters are discussed in the fuel-supply system chapter.

Three needle/seat assemblies: Pointed needle with Viton tip (Viton tips may be any color), pointed steel-tip needle and ball-tip needle with float hook.

Two spring-loaded needle/seat assemblies. These are used to provide cushioning of the needle when the float bounces the needle upward against the seat. Hooked needle has captured spring and ball end in a hollow needle.

Cutaway race-type bowl shows adjustable needle/seat assembly 1, adjusting nut 2 and lock screw 3. Sintered-bronze filter 4 in inlet has spring 5 to allow fuel bypass if filter plugs up. Fuel-inlet nut 6 can be swapped with plug 7 on some race bowls to allow plumbing to opposite side. Inside view shows half of float 8 and segment of mounting bracket 9.

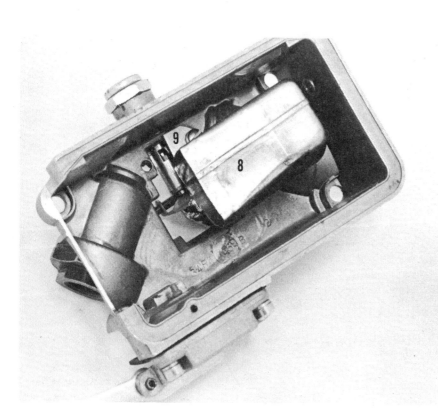

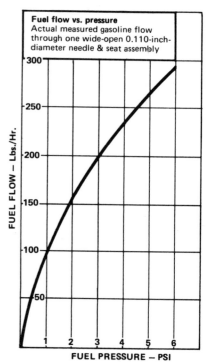

Fuel flow vs. pressure
Actual measured gasoline flow through one wide-open 0.110-inch-diameter needle & seat assembly

FUEL FLOW – Lbs./Hr.

FUEL PRESSURE – PSI

Actual measured gasoline flow through one wide-open 0.110-inch-diameter window-type needle/seat assembly. Float held at bottom of bowl.

SUMMARY

Before leaving the inlet system, we should list the items which can affect fuel level:

1. Inlet-pressure changes (fuel-pump output pressure)

2. Float bumper spring

3. Inlet seat size

4. Float density

5. Float size

6. Float-lever length

7. Fuel density (density of gasoline remains fairly constant, so this is not usually a problem).

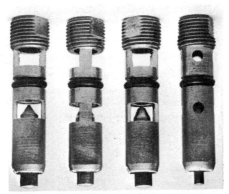

Round-hole needle/seat assembly is used for low-flow applications. "Picture-window" assemblies are common where higher flow is required. O-ring seals inlet side from fuel bowl. Threaded portion provides for fuel-level adjustment when used with adjusting nut and locking screw.

Viton-tipped needle is captured in seat by cap at left so that needle/seat can be replaced as an assembly.

FUEL FLOW vs. FUEL PRESSURE FOR VARIOUS SIZE NEEDLES & SEATS

Dia. of Needle Seat	Type	Fuel Flow @ 2 P.S.I lbs./hr.	Fuel Flow @ 4 P.S.I. lbs./hr.	Fuel Flow @ 6 P.S.I. lbs./hr.
0.082"	Holes	106	153	204
0.097"	Holes	121	174	225
0.101"	Holes	138	194	254
0.110"	Holes	153	230	275
0.110"	Windows	160	232	295
0.120	Windows	167	236	305

NOTE: Checked with needle & seat assembly installed with float held at bottom of bowl.

Holley Tests
October, 1971

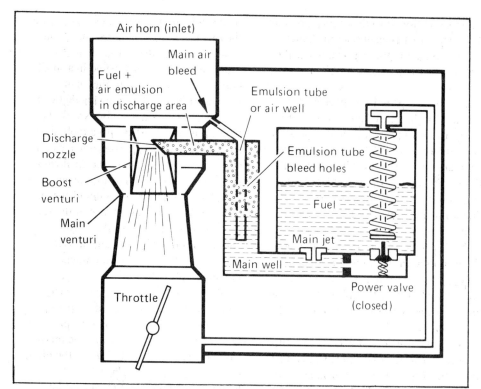

The main venturi's reduced diameter speeds up air flow. Reduced pressure in the venturi area allows the air stream to pick up fuel from the bowl through sometimes-complicated passages. A main jet in the fuel path limits fuel flow.

Boost venturis may also have wings or tabs (arrows) to make a carburetor work on a specific engine/manifold combination. Picture how these work by touching the edge of water streaming from a faucet. Mixture is deflected in the same way.

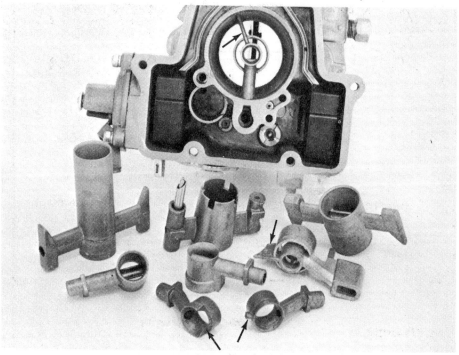

Main System

The main metering system supplies fuel/air mixture to the engine for cruising speeds and above. During the time engine speed or air flow is increasing to a point where the main metering system begins to operate, fuel is fed by the idle and accelerating-pump systems. Under conditions of high load, when the engine must produce full power, added fuel comes from the power system.

Many people believe the throttle controls the *volume* of fuel/air mixture being pumped into the engine. This is not the case. Piston displacement never changes, so the *volume* of air pulled into the engine is constant for any given speed.

The throttle controls the *density* or mass flow of the air pumped into the engine by the action of the pistons: Least *density* of charge is available at idle, highest *density* is at wide-open throttle. A dense charge has more air mass, hence higher compression and burning pressures can be developed for higher power output. Thus, the throttle controls engine speed and power output by varying the *charge density* supplied to the engine.

Now that we've discussed the throttle, let's proceed to the heart of the carburetor and the key to its simplicity, the *venturi*. It is the one part that really makes a carburetor function, so it is important to understand how the venturi operates before getting deeper into how the main metering system works.

The venturi and its principle of operation are named after G. B. Venturi, an Italian physicist (1746-1822) who discovered: When air flows through a constricted tube, flow is fastest and the pressure lowest at the point of maximum constriction.

In the internal-combustion engine, a partial vacuum is created in the cylinders by the downward strokes of the pistons. Because atmospheric pressure is higher than the reduced pressure in the cylinders, air rushes through the carburetor and into the cylinders to fill the vacuum. On its way to the cylinders, the air passes through the venturi.

The venturi is a smooth-surfaced restriction in the path of the incoming air. It "necks-down" the inrushing air column, then allows it to widen back to the throttle-bore diameter. Air is rushing in with a certain pressure. To get through the necked-down area (venturi), it must speed up, reducing the pressure inside of the venturi. A gentle diverging section is used to recover as much of the pressure as possible.

The venturi is the controlling factor in the carburetor because fuel discharges into the venturi at the point of lowest pressure (greatest vacuum). This minimum-pressure point applies a *signal* to the main-metering system.

The pressure drop or vacuum "signal" is measured at the discharge nozzle in the venturi. Because the fuel bowl is maintained at near-atmospheric pressure by the vent system, fuel flows through the main jet and into the low-pressure or vacuum area in the venturi.

Pressure drop (vacuum) at the venturi varies with engine speed and throttle position, increasing with engine RPM. Wide-open throttle and peak RPM give the highest flow and the highest pressure difference between the fuel bowl and a discharge nozzle in the venturi, thus the

highest fuel flow into the engine. Pressure difference (ΔP) as engine speed changes is approximately proportional to the difference in velocity squared (ΔV^2), $\Delta P \sim V^2$.

Pressure drop in the venturi also depends on the size of the venturi itself. A small venturi provides a higher pressure drop at any given RPM and throttle opening than a large venturi will provide. The Design Features portion of this section tells you more about this important consideration and how it relates to performance.

No fuel issues from the discharge nozzle until flow through the venturi and hence, pressure drop, is sufficient to offset the level or "head" difference between the discharge nozzle and the lower level of fuel in the bowl.

Once main-system flow is started, fuel is metered (measured) through a main jet in the fuel bowl. From the main jet, fuel passes into a main well. As fuel passes up through this main well, air from a main air or "high-speed" bleed is added to pre-atomize or emulsify the fuel into a light,

frothy fuel/air mixture which issues from the discharge nozzle into the air stream flowing through the venturi. The discharge nozzle is often located in a small boost venturi centered in the main venturi (described in the following section).

There are two main reasons why the liquid fuel is converted into a fuel/air emulsion. First, it vaporizes much easier when it is discharged into the air flowing through the venturi. Second, the emulsion has a lighter viscosity than liquid fuel and responds faster to any change in the venturi vacuum (signal from the venturi applied through the discharge nozzle). It will start to flow sooner and quicker than purely liquid fuel.

The strong signal from the discharge nozzle is bled off or reduced by the main air bleed so there is less effective pressure difference to cause fuel flow. The mixture will become leaner as the size of the bleed is increased. Decreasing the bleed size increases the pressure drop across the main jet to pull more fuel through the main system, giving a richer mixture. Main-air-bleed changes affect the entire range

Typical vacuums inside a carburetor air section. Slight vacuum at inlet represents drop across the air cleaner. Gage at large venturi throat shows higher vacuum because air is still at relatively high velocity. Vacuum returns almost to the inlet value just before the throttle plate. Throttled carburetor shows very high manifold vacuum because a large pressure drop occurs across the partially-opened throttle. Wide-open carburetor's low manifold vacuum indicates a heavy-load/dense-charge situation. In both cases, the highest carburetor vacuum is at the boost-venturi throat which supplies the signal to the main system.

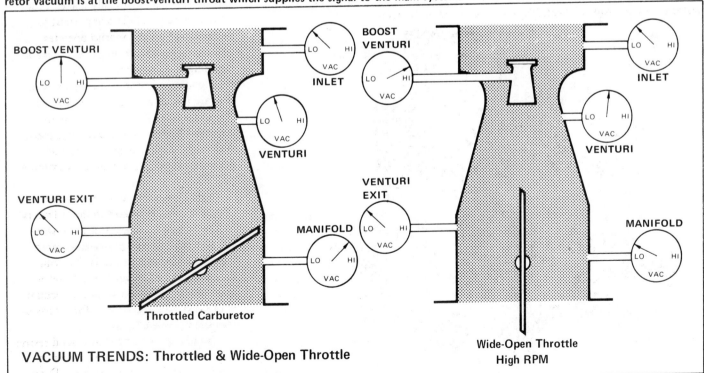

VACUUM TRENDS: Throttled & Wide-Open Throttle

Throttled Carburetor

Wide-Open Throttle
High RPM

Air from the main air bleed can be introduced into the main well in several ways. Sometimes it is brought in through the center of an emulsion tube. Separate emulsion tubes shown here are from a 2110 Bug Spray (right) and a 5200 (left). In carburetors with removable metering bodies, bleed air is often introduced through tiny holes in an air well paralleling the main well. This well is on the surface assembled toward the carburetor body. Arrows indicate bleed holes which are necked down to tiny holes 0.026-inch diameter or so.

of main-metering-system operation. Holley establishes the main-air-bleed size for each carburetor to work correctly over that carburetor's air-flow range. Changes in main-air-bleed sizes are rarely necessary or advisable; calibration changes are easy to make by changing main metering jets.

The main air bleed also acts as an anti-siphon or siphon-breaker so fuel does not continue to dribble into the venturi after air flow is reduced or stopped.

DESIGN FEATURES/Main System

Throttle—The throttle shaft is offset slightly—about 0.020 inch on primaries—about 0.060 on secondaries—in the throttle bore so one side of the throttle has a larger area to cause self-closing. There are two reasons for this feature. First, idle-return consistency is greatly aided by the sizable closing force generated when the manifold vacuum is high—as at idle. Second, it is a safety measure to guard against overspeeding the engine if it is started without installing the linkage or throttle-return spring because air flow past the throttles will tend to cause them to close.

Throttles are seldom closed tightly against the throttle bore; instead, they are factory set against a stop to provide a closed-throttle air flow specified for that particular carburetor list number.
Venturi—A few comments about venturi design. The most efficient (or ideal) venturi which creates the maximum pressure drop with the minimum flow loss requires a 20° entry angle and a diverging section with a 7° to 11° included angle on the "tail." The "ideal" venturi will not always fit into a carburetor which will fit under the hood of the average automobile. Holley uses a radiused entry to the venturi because it is less critical to production variations. Designers try to keep as close as possible to the "ideal" venturi entry and exit angles within the limitations imposed by engine position and hood heights.

Although the theoretical low-pressure point and point of highest velocity would be expected at the minimum diameter of the venturi, friction causes the point to occur about 0.030 inch below the smallest diameter. This low-pressure, high-velocity point is called the *vena contracta*. The center of the discharge nozzle or the *tail* of the boost venturi is located at this point.

A venturi allows a much greater metering signal than a straight tube and has a minimum loss of air flow. This is because the trailing edge of the venturi conforms to the normal air stream and therefore recovers most of the pressure drop so a greater mass of air is available for the engine.

The venturi is a very efficient air-metering device because of the high signal levels it provides in conjunction with minimum pressure loss.

The venturi provides a way to supply fuel in the correct proportion to the mass of air rushing into the engine. Its size, as mentioned in the description of how it works, affects the pressure drop available to operate the main-metering system. The smaller the venturi, the greater the pressure drop, the sooner the main system will be brought into operation, and the better the mixing of fuel with air will be. The RPM at which main system "spill-over" or "pull-over" starts is affected by the size of the engine which is pulling air through the carburetor.

In plainer words: If you keep engine size constant, a larger carburetor requires a higher RPM to bring the main system into operation; a smaller carburetor—a lower RPM. But, the venturi size also controls the amount of air available at wide-open throttle (WOT) for engine operation. If the venturi is too small, top-end HP will be reduced, even though the carburetor provides very good fuel/air mixing at cruising speeds. For this reason automobile manufacturers typically compromise—by using carburetors with smaller than optimum (maximum power) air-flow capacities. They do this for very good reasons, even if you—a performance enthusiast—may feel cheated. Good fuel atomization and vaporization promotes good distribution and improves smooth running and economy at around-town and cruising speeds.

Manufacturers offering more performance typically supply a carburetor with a secondary system which retains the advantages of a small primary venturi with the capability of higher air flow for top-end power. Secondary systems are described in the last section of this chapter.
Boost Venturi—Boost venturis act exactly the same as the larger venturi, but supply

a stronger signal to the discharge nozzle because boost-venturi velocity is higher than that in the main venturi. Boost venturis are *signal amplifiers*.

By increasing the available signal for main-system operation, the boost venturi allows the carburetor to work well at lower speeds—and therefore lower air flows—than would otherwise be the case. This is especially helpful in a performance-type carburetor because the boost venturi does not seriously affect the air-flow capacity of the carburetor.

The tail of the boost venturi discharges at the low-pressure point *(vena contracta)* in the main venturi. Thus, the boost-venturi air flow is accelerated to a higher velocity because it "sees" a greater pressure differential than the main venturi.

Air and fuel emerging from the boost venturi are traveling faster than the surrounding air, therefore, a shearing effect takes place between the two air streams. This improves atomization of the fuel.

Boost venturis also aid fuel distribution because the ring of air which flows between the two venturis directs the charge to the center of the air stream, helping keep some of the wet fuel/air mixture off of the carburetor wall below the venturi so that more of the fuel/air mixture reaches the manifold hot spot for improved vaporization. In addition, tabs, bars, wings and other devices are sometimes needed to provide correct directional effects for good cylinder-to-cylinder distribution with a particular manifold. All such development work is accomplished on the dynamometer during design and development of the carburetor/manifold combinaiton.

Boost venturis allow using a much shorter main venturi so the carburetor can be made short enough to fit under the hood of an automobile. Carburetor designers could achieve essentially the same results with a long (ideal) venturi as they can with one or more boost venturis "stacked" in the main venturi, but it is very tough to build a carburetor which allows the 20° entry to the venturi and a 7° to 11° "tail," and at the same time get venturi size down to the required diameter for adequate signal. Some Holley carburetors, especially very short ones, use as many as two booster venturis to get adequate signal for main system operation.

Main Jets—These metering orifices are used to control fuel flow into the metering system. They are rated in flow capacity and are removable for calibration purposes. It is commonly thought that main jets are selected solely on a trial-and-error basis, but this is not the case. The accompanying material clearly shows the relationship of main-jet size to venturi size. For a given venturi size, a small range of main jets will cover all conditions. However, final selection of the correct jet for the application will have to be done by testing because design and operational variations (climate, altitude and temperature) affect jet-size requirements.

There is a basic misconception about jets—that size alone determines their flow characteristics. This is not the case because the shape of the jet entry and exit—as well as the finish—affects flow. Holley checks each jet on a flow tester and grades it according to flow. This flow rate is compared to a master chart and a number is stamped on the jet to indicate its flow. The tolerance range used for each size explains why a 66 jet may not seem to give a richer mixture than a 65—if the 65 is on the "high" side of its allowable tolerance and the 66 is on its "low" side tolerance limit—the two jets may flow very close to the same amount of fuel. The tolerance range for a 65 ranges from 351.5 to 362.0 cubic centimeters per minute at a specified head with a given test fuel. The 66 ranges from 368.5 to 379.5 cc's.

In each of these cases, there is a 3% flow range within a jet size, and about 4.5% difference in flow between average jets in the two sizes. In 1975 Holley developed a new close-limit series of main jets in the standard size range of 30 to 74. This new series was developed so flow could be more closely tailored to fit within emission requirements. These jets still use the first two numbers, such as 65 or 66, but a third number is added to indicate whether the jet flows toward the lean side (651) in the middle (652), or rich (653). There is approximately 1.5% difference in flow between each of the three jets, or a flow range of 4.5%. There can only be a 1.5% difference in flow between two jets with the same flow marking, as opposed to the possibility of a variation

of as much as 3% flow difference between jets with the same markings in the old two-digit numbering system.

The new jet series has the same brass color or aluminum color as was previously used.

Holley offers the close-limit jets in the mean sizes only, that is, 352, 262, etc. You may have a carburetor with a 351 or a 353 in it, but these jets are only available in the assembly plant where the carburetor is made and flow checked. Close-limit jets offer a good way to accomplish fine tuning, provided you can use jets ranging from 35 to 74 (352 through 742). Order these jets as 22BP-120-XX2.

Some jets used in the Models 5200 and 5210 were marked to indicate diameter of the jet opening in millimeters: 130 = 1.3 mm. As of 1976, a system of jet marking relating directly to the flow in cubic centimeters per minute was adopted for use in the Models 5200, 5210 and 4360 carburetors. These jets are dyed green for identification so you can identify that they are marked according to flow. A jet marked 357 has a tolerance ranging from 354.3 to 359.7cc, or 1.5% maximum difference between two jets with the same markings. The flow rating is made at 50 centimeters head. Order these jets as 22BP-130A-XXX. Consult the current Holley Performance Parts Catalog for available sizes.

Drilling out jets to change their size is never recommended because this always destroys the entry and exit features to a certain degree, and may introduce a swirl pattern, even if the drill is held in a pin vise and turned by hand. You cannot be sure of the flow characteristics of a jet that you modify by drilling—unless you can get that jet back on a flow machine to compare it with a standard jet.

Main Air Bleed—All Holley carburetors except Models 5200 and 5210 are equipped with built-in fixed-dimension air bleeds. A flat surface in the inlet horn is typically used for mounting these bleeds so that they will see total pressure and will not be affected by air-flow variations.

Relationship between Main Jet & Venturi Size—By using some standard equations for fuel flow and air flow, we can show how the fuel-air ratio is established in a carburetor—within certain limits.

The mass of fuel flowing through the main jet can be calculated by:

(1) $\ M_f = A_{mj} \times C_{mj} \sqrt{2g_c\, \gamma_f\, \Delta P_{mj}}$

where

M_f = mass fuel flow
A_{mj} = cross-sectional area of main jet
C_{mj} = coefficient of discharge of main jet
g_c = dimensional constant—acceleration of gravity
γ_f = density of fuel
ΔP_{mj} = pressure drop across main jet

Lumping all the constants together:

$$C_{mj} \times \sqrt{2g_c} = K_f$$

Therefore equation (1) can be restated as:

(2) $\ M_f = A_{mj} \times K_f \times \sqrt{\gamma_f} \times \sqrt{\Delta P_{mj}}$

This gives the mass flow of an incompressible fluid through a restricted orifice—the main jet. When the reduction in air pressure through the venturi is small, air also behaves nearly the same as an incompressible fluid, so an equation of the same form can be used to state the mass of air flowing through the venturi restriction:

(3) $\ M_a = A_v \times C_v \times \sqrt{2g_c\, \gamma_a\, \Delta P_v}$

where

M_a = mass flow of air
A_v = cross-sectional area of the venturi
C_v = coefficient of discharge of the venturi
γ_a = density of air
ΔP_v = pressure drop from the venturi inlet to throat

This can also be restated by combining constants:

(4) $\ C_v \times \sqrt{2g_c} = K_a$

The equation for mass air flow can then be written:

(5) $\ M_a = A_v \times K_a \times \sqrt{\gamma_a} \times \sqrt{\Delta P_v}$

Fuel-air ratio is the mass of fuel divided by the mass of air:

$$\frac{M_f}{M_a}$$

This division can be done using equation (2) for mass of fuel and equation (5) for mass of air flow:

(6) $\ \dfrac{M_f}{M_a} = \dfrac{A_{mj}}{A_v} \times \dfrac{K_f}{K_a} \times \dfrac{\sqrt{\gamma_f}}{\sqrt{\gamma_a}} \times \dfrac{\sqrt{\Delta P_{mj}}}{\sqrt{\Delta P_v}}$

Now this complicated-looking expression starts falling apart and becomes utterly simple.

First, K_f and K_a can be disregarded because they are collections of constant numbers which never change.

Because we assumed both gasoline and air are incompressible, their densities don't change and the ratio of their densities becomes another constant we can disregard.

In the carburetor, one side of the venturi passage and one side of the main jet are both at atmospheric pressure. The other sides are both at the lowest pressure caused by the venturi, so the pressure differential across the main jet (ΔP_{mj}) is equal to the pressure differential across the venturi (ΔP_v). Their ratio is one and can be disregarded also.

After throwing out all the constants, we no longer have an equation but we have a proportionality which tells us:

$\dfrac{M_f}{M_a}$ is proportional to $\dfrac{A_{mj}}{A_v}$

or, $\quad \dfrac{M_f}{M_a} \sim \dfrac{A_{mj}}{A_v}$

Because the mass of fuel divided by the mass of air is the fuel-air ratio,

$$F/A \sim \frac{A_{mj}}{A_v}$$

This simply means that over the flow ranges considered, if the relationship between the venturi size and the main jet remains the same, a constant fuel/air ratio will be maintained.

Unfortunately air *is* compressible and as pressure drops are increased our relationship does not hold. It is a very good approximation when dealing with high-performance/low-restriction carburetion.

What really happens is that as air flow approaches the critical value (sonic velocity) and ΔP increases, γ_a gets lower and lower. Looking at expression (6) we can see that the fuel/air ratio will increase.

These main jets all look alike except for the markings. The 50 is the standard jet. 501, 502 and 503 are a new close-limit series with only 1.5% flow range difference (maximum) between any jet with the same marking. Two-digit (50) series jets had 3% possible variation between any two jets with the same marking.

Power System

When the engine is called upon to produce power in excess of normal cruising requirements, the carburetor has to provide a richer mixture, as discussed in the fuel-requirements section. Added fuel for power operation is supplied by the power system under control of manifold vacuum. Manifold vacuum accurately indicates the load on the engine. It is usually strongest at idle. As load on the engine increases, the throttle valve must be opened wider to maintain a given speed, thereby offering less restriction to the air entering the intake manifold and reducing the manifold vacuum.

A vacuum passage in the carburetor applies manifold vacuum to a power valve piston or diaphragm. At idle or normal cruising load conditions, manifold vacuum acting against a spring holds the valve closed. As high power demands load the engine, and manifold vacuum drops below a pre-set point—usually at about 6 inches of mercury (Hg)—the power-valve spring overcomes manifold vacuum and opens the power valve. Fuel flows through the power valve and through a power-valve restriction to join fuel already flowing through the main metering system from the main jet, thereby richening the mixture. The power valve can be considered as a "switch" which turns on extra fuel to change from an economical cruising fuel/air mixture to a power mixture. The power-valve channel restriction (PVCR) controls the amount of enrichment. The power valve itself is merely a switch operated by manifold vacuum. It is designed to operate (add more fuel) at a given load.

When engine power demands are reduced, increasing manifold vacuum acts on the diaphragm or piston to overcome the power-valve-spring tension, closing the power valve and shutting off the added fuel supply.

DESIGN FEATURES/Power System
Power Valves—Holley high-performance carburetors usually have single-stage power valves. Carburetors for street use, especially those for 1973 and later models with exhaust-gas recirculation, have two-stage power valves. Two-stage power valves are discussed later.

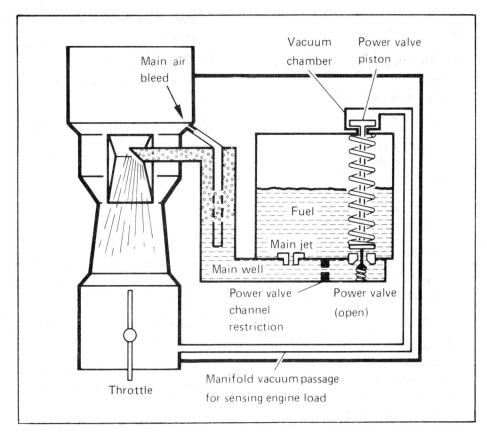

Single-stage power valves of the screw-in type are used in the 2300, 4150, 4160, 4165, 4175 and 4500 series. These are available with different flow areas and opening points. The flow area used must always be larger than the combined areas of the power-valve channel restrictions (PVCR's).

Opening points are available in increments from 1.0 to 10.5 inches Hg. The last digits in the part number indicate the opening point in inches of mercury for that power valve when a decimal point is placed ahead of the last digit. This number is also stamped on the valve. Examples:

25BP-591-105A opens at 10.5 in. Hg
25BP-591A-25 opens at 2.5 in. Hg

The power-valve opening point is another variable the carburetor designer uses to arrive at the best compromise between economy, exhaust emissions and drivability. On normal replacement carburetors, the valve may be opened quite late to allow a particular engine to meet emission requirements, as well as maintain a broad economy range. Racing engines, on the other hand, have wild manifold-vacuum fluctuations at idle and low speeds. The user must install a power valve which will not open and close in response to these variations which are caused by valve timing and not by throttle position or engine load. Selecting power valves to meet these special requirements is explained in the performance-tuning chapter, page 131.

Most Holley two- and four-barrel performance-type carburetors use 25BP-591A-XX "picture-window" type power valves with flow capacity sufficient for two 0.095-inch PVCR's.

One picture-window power valve, the 25BP-595A-XX used in one Model 4500 and in the big 6425 two-barrel, flows fuel to handle two 0.128-inch PVCR's. Because of its size, the valve stem cannot be piloted and the power valve can be damaged by dirt in the fuel. Don't use the -595's unless you have a real need for them.

Six-hole power valves, 25BP-398A-XX and 25BP-400A-XX have also been used on older high-performance Holleys. It was originally thought that these flowed more fuel than the 25BP-237A-XX valves with four 0.073-inch holes, but tests show the six-hole flows the same as the four-hole. Either valve flows fuel enough for two 0.067-inch PVCR's. Incidentally, the six 0.104-inch holes seriously weaken the -398 and -400 valves, so great care must be used in installation and removal to avoid twisting the valve apart at the holes.

Screw-in power valves may be stamped with additional numbers, such as A5. This mark indicates the month and year in which the valve was made. A is January and 5 is 1975. C6 indicates March, 1976.

Two-stage power valves are used on normal replacement carburetors to allow a particular engine to pass emission tests. These valves are also used on some performance carburetors, as mentioned earlier. Staged power valves open partially at one vacuum level, then fully open when manifold vacuum falls to a lower level. The first stage in these valves opens a metering orifice smaller than the PVCR.

With the rising interest in gasoline economy, Holley now offers the two-stage power valves as replacement parts. These provide improved drivability in vehicles with relatively low power-to-weight ratios. Although Holley cannot guarantee economy improvements, some benefit should be realized if the two-stage power valves are applied as follows:
1. Do not use the valves in any vehicle which will be used on the drag strip.
2. The best economy improvement will be realized in recreational vehicles equipped with the R-6619 and R-6909 carburetors, and in some station wagons and heavy sedans using the Model 4165 spread-bore carburetors. Use the valve only on the primary side in the Model 4165. Best results will be obtained in stop-and-go driving, or where the vehicle is operated in rolling or mountainous terrain.
3. In other applications, some fuel economy may be lost. Typical two-stage power valves operate as follows: 25BP-475A-12 opens the 1st stage at 12 inches Hg, the 2nd stage at 6 inches Hg. For high-altitude applications above 4,000 feet, the 25BP-475-25 opens the 1st stage at 10 inches Hg, the 2nd at 5 inches Hg.

Piston-type power valves are used in Models 1940, 1945, 2210, 2245, 1901 and 4360.

The spring force is factory-set to open at a specified manifold vacuum. This setting can be changed by adding or deleting shims from the spring at the foot of the piston stem. The greater the number of shims, the higher the manifold vacuum at which the valve will open. If the spring is shortened, or shims removed, the valve will open at a lower manifold vacuum.

New power valve at right introduced in 1974 discharges through cast slots for increased flow and cross-sectional strength. The new valve is stronger than the drilled type and less likely to break off in installation or removal. It is interchangeable with all older units, but gasket 8R-1597 with no protrusions on the inside diameter must be used. The valve is offered in two sizes: 25R-591-A for all power-valve-channel restrictions up to 0.090 inch; 25R-595-A for rare calibrations using larger PVCR's.

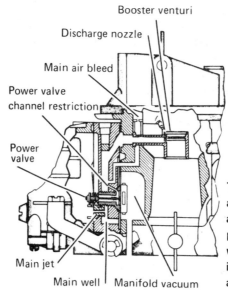

Booster venturi
Discharge nozzle
Main air bleed
Power valve channel restriction
Power valve
Main jet
Main well
Manifold vacuum

Typical vacuum and fuel-passage routing as used in Holley carburetors with removable fuel bowls. These are diaphragm-type power valves. Using a secondary power valve allows using a smaller main jet to improve fuel handling during deceleration and braking.

Three types of Holley power valves. Top is a typical single-stage screw-in type used on the Model 2300, 4150, 4160, 4165, 4175, 4500 and 3150 carburetors. Center is a gradient power valve used on Model 2245 and 1945. *Gradient* means the piston operates over a greater-than-normal range of manifold vacuum and meters proportionally with a tapered valve. The piston and spring are retained in the air horn by the clip. Valve, seat and spring are screwed into the main body. Hardware at bottom is from Models 5200 and 5210: Diaphragm and spring assembly is attached to the air horn by three screws, seat, valve and spring go into the main body.

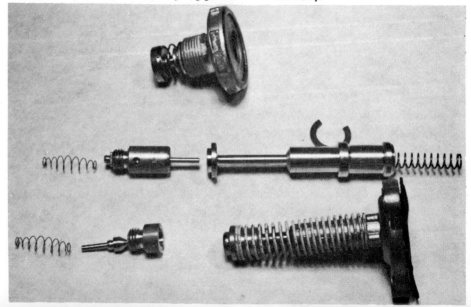

Numbers stamped into flat on screw-in power valves indicate manifold vacuum at which valve opens: 4.5, 6.5 and 6.5 inches Hg. Left has four 0.073-inch holes, second valve has six 0.076-inch holes and third valve has six 0.104-inch holes. Fourth valve shows opening to vacuum-controlled diaphragm. Several fuel-inlet holes in the power valve ensure that the valve itself does not become a restriction. The only time that metering is accomplished in the power valve is in the first stage of two-stage power valves. The power-valve channel restriction is sized to supply the additional fuel required for power enrichment.

Three power valves. Small unit at left is operated by piston rod. Center unit is two-stage type for 4150/60, 4165, 2300 models. Unit at right fits same carburetors, is inverted power valve which is normally open and closes when manifold vacuum drops. Number stamped on flat of this valve was 30, indicating that the valve closed at 3.0 inches Hg manifold vacuum.

Idle System

Idling requires richer mixtures than part-throttle operation. Unless the idle mixture is richer, slow and irregular combustion will occur due to the high dilution of the charge by residual exhaust gases which exist at idle vacuums. This is described in the engine requirements section.

DESCRIPTION OF OPERATION

The idle system supplies fuel at idle and low speeds. The idle system has to keep the engine running, even when accessory loads are applied to the engine. These include alternator, air conditioning, and power-steering pump. The idle system also has to keep the engine running against the load imposed by placing an automatic transmission in one of the operating ranges (Low, Drive, Reverse).

At idle and low speeds, not enough air is drawn through the venturi to cause the main metering system to operate. Intake-manifold vacuum is high because of the great restriction to the air flow by the nearly closed throttle valve. This high vacuum provides the pressure differential for idle-system operation.

Perhaps the easiest way to understand how the idle system works is to look at it as a tiny main metering system. Because air enters the idle system through the idle air bleed, think of the bleed as the venturi. The amount of flow through this system depends on the area of the transfer slot and/or discharge hole exposed by the main throttle plate and the position of the idle-mixture screw. Consider the discharge hole or slot, and idle screw as forming the "idle-system throttle."

Backing the idle screw out or opening the throttle to expose more of the slot to manifold vacuum opens the idle system "throttle" so more air flows through the system. The drop across the idle-air bleed increases, bringing along more fuel from the idle well so the mixture stays at the desired fuel/air ratio. Fuel flows through the main jet, then into a vertical idle well, past the idle-feed restriction and is then mixed with air from the idle air bleed. Sometimes the idle restriction is placed at the bottom of the idle well. This fuel/air mixture is lifted up and across to another vertical passage. Fuel-air mixture flows down this second vertical passage

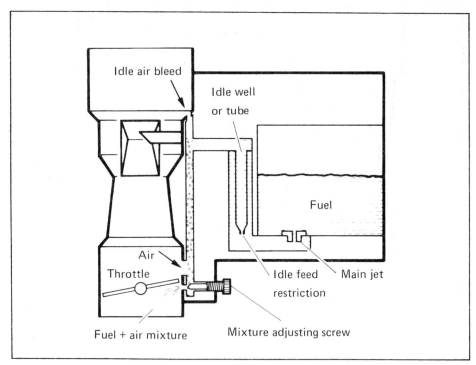

Idle system operates like the main system, only on a smaller scale.

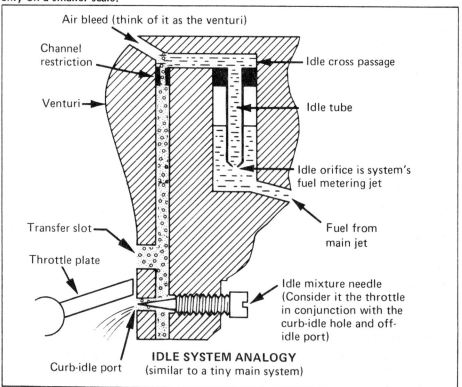

IDLE SYSTEM ANALOGY
(similar to a tiny main system)

and then branches in two directions; through the idle-discharge passage into the throttle bore *below* the throttle valve, and to an idle-transfer slot (or holes) located just above the throttle valve.

As the throttle valve is opened and engine speed increases, air flow through the venturi is increased so that operation of the main metering system begins dis-charging fuel through the discharge nozzle in the booster venturi. Flow from the idle system tapers off as the main system starts to discharge fuel. The two systems are designed to provide smooth gradual transition from idle to cruising speeds *when the carburetor capacity is correctly matched to engine displacement.*

In normal driving, flow swings quickly

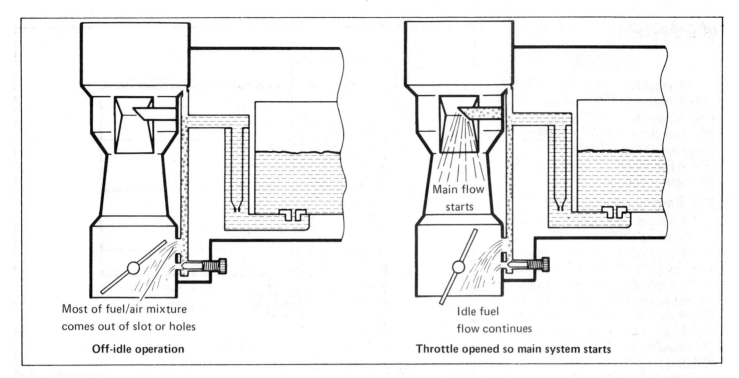

Most of fuel/air mixture
comes out of slot or holes

Off-idle operation

Main flow
starts

Idle fuel
flow continues

Throttle opened so main system starts

back and forth between idle and main operation as the vehicle is accelerated, slowed by closing the throttle, idled at stop, and then reaccelerated.

When the throttle is closed, high vacuum below the throttle plate draws air and fuel through the idle system and out the idle-discharge port. When the throttle is partially open, vacuum below the throttle plate is reduced and flow through the idle system is reduced, but some flow is out the idle port and some is out the idle-transfer port. When the throttle is fully open, there is not enough vacuum at the idle port to draw any fuel-air mixture through the idle system. Operation of the idle system gradually ceases as the throttle is moved from idle to cruising. Naturally, flow through the main-metering system increases as idle flow decreases, so the transition is smooth—*if* the right carburetor is chosen for the engine.

DESIGN FEATURES/Idle System Throttle Stop—Seating the throttle against a stop instead of closing it fully against the throttle bore ensures against the throttle sticking in the bore and makes the idle system less sensitive to mixture adjustments. Holley carburetors are factory-set to an idle air-flow specification and they should not be readjusted to seat the throttle in the bore (especially true on diaphragm-operated secondaries).

Idle-Air-Bleed Size—Increasing the idle-air-bleed size reduces the pressure drop

across the bleed, decreasing the amount of fuel which can be pulled over from the idle well. Thus, increasing idle-air-bleed size leans the idle mixture, even if the idle-feed restriction is left constant. Conversely, decreasing the size of the idle-air bleed increases the amount of pressure drop which can be obtained in the system and richens the idle mixture.

Idle-Speed Setting—Before emission-control requirements became important, idle setting was typically the slowest speed at which the engine would keep running smoothly. Emission requirements have made higher idle speeds necessary. A higher idle speed reduces some of the exhaust-gas dilution which occurs at lower idle speeds so leaner idle mixtures can be used without misfiring.

Cars with engines designed to pass emission requirements; i.e., production-type automobiles, are typically set for lean best idle at specified RPM and a subsequent reduction in idle speed by leaning the mixture still further—as stated on a label in the engine compartment.

The manufacturers have carefully correlated this lean best idle and subsequent idle drop off as that which provides the required carbon monoxide percentage (CO%) to pass the requirement. From this lean-best-idle point, and at this specified RPM figure, each of the idle-mixture-adjustment screws are turned in to provide a specified RPM drop off. Where

there are two screws, each is adjusted to provide 1/2 of the specified idle drop off. Fuel/air ratios are usually stated for use when an analyzer is available.

Older non-emission-controlled cars and racing cars are typically idle set for the desired idle RPM and best manifold vacuum. This is not a minimum-emission setting, however.

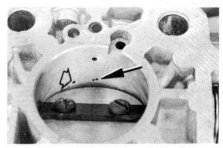

Some idle transfer circuits (also called progression or off-idle) use a slot as shown above. Others use a series of holes. Curb idle is always the hole nearest the flange. Outline arrows indicate spark ports. This view is from the manifold surface.

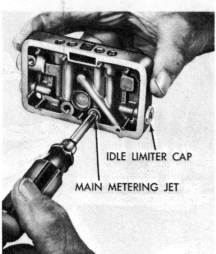

IDLE LIMITER CAP

MAIN METERING JET

Idle limiters applied to idle-mixture adjustment screws hold idle-system adjustments within a narrow band which provides specification emission levels for a given engine and carburetor combination. Limiter caps either cover the screw access openings (as at right) or fit on the screws (left). In either case, no further adjustment can be made without destroying the cap/s.

Idle Limiter—An idle limiter cap limits idle-mixture-screw adjustment to approximately 3/4 turn. Applied after the desired idle mixture has been set, this limiter prevents easy tampering with the idle-mixture adjustment. This factory setting is made when the carburetor is flow checked.

The limiter is constructed so removing it requires destroying the cap, thereby showing instantly that the carburetor has been readjusted and may not be providing required emission performance.

Intermediate Idle System—In some of the largest Holley high-performance carburetors, such as the Model 4500, List 6214 and the 6464, an intermediate idle system discharges through a tube into the trailing edge of the venturi. This extra idle system provides additional transfer fuel between the idle- and main-system operation. Opening the throttle past the usual idle-system transfer slot greatly reduces the manifold vacuum which was being applied to the idle system. But, in these large-venturi carburetors, velocity through the venturi is not sufficient to establish main-system flow at this time. A giant flat spot would be present without this intermediate idle system.

The discharge tube for the intermediate idle system is located above the transfer slot at the main venturi tail. Operation of the intermediate system starts as the throttle moves past the intermediate discharge. The system flows continuously beyond this point.

The intermediate idle system has its own air idle-air bleed and idle-feed restriction. There are no adjustments. However, the bowl-fed system continues feeding

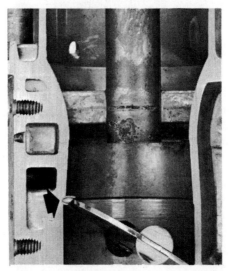

Cutaway 4500 shows intermediate idle-system discharge (arrow). System is fed from its own restriction in fuel bowl.

fuel after the main system starts because the discharge tube is at a lower pressure than the fuel bowl.

Auxiliary air bleeds—Auxiliary air bleeds are sometimes used in the idle system. Although these usually add air to the idle system downstream from the traditional idle air bleed, they act in parallel with the idle air bleed.

Accelerator-Pump System

The accelerator pump has three functions:

1. To make up for the fuel that condenses onto the manifold surfaces when the throttle is opened suddenly.

2. To make up for the lag in fuel delivery when the throttle is opened suddenly, which allows more air to rush in.

3. To act as a mechanical-injection system to supply fuel before the main system starts.

As the throttle is opened quickly, intake manifold vacuum instantly drops, moving the pressure toward atmospheric. A high manifold vacuum tends to keep the mixture well vaporized. As the pressure rises toward atmospheric, fuel drops out of the vapor, condensing into puddles and wet spots of liquid on the walls and floor of the intake manifold. Thus, the mixture which is available for the cylinders is instantly leaned out and the engine hesitates or stumbles unless more fuel is immediately added to replace that which has fallen out of the mixture onto the manifold surfaces. This is especially important with big-port manifolds or manifolds with large plenums (volume below carburetor) because these have more surface area onto which fuel can condense.

Making up for condensed fuel loss to the manifold and taking care of fuel-delivery lagging behind increasing air flow are both important, but the relative importance of the two has not been established.

A third function of the accelerator-pump system is supplying fuel when the throttle is quickly opened past the point where the idle-transfer system would have supplied fuel until the main system could begin its normal operation. In this case, the accelerator pump supplies the required fuel until the main system starts flowing. During the low-flow, low-vacuum period the accelerator pump injects fuel under pressure into the throttle bore. Duration of accelerator-pump operation must be carefully engineered to provide a "cover up" of sufficient length to allow main-system flow to be established so good vaporization will be ensured and correct fuel/air ratio reestablished.

The accelerator pump operates when the pump-operating lever is actuated by throttle movement. As the throttle opens, the pump linkage operates a pump diaphragm or plunger. Pressure in the pump forces the pump-inlet ball or valve onto its seat so fuel will not escape from the pump into the fuel bowl. Pressure also raises the discharge needle or ball off its seat so fuel is discharged through a "shooter" into the venturi.

As the throttle is moved toward the closed position, the linkage returns to its original position and the pump-inlet ball or valve is moved off its seat to allow the pump to refill from the bowl. As the piston or diaphragm is positively pulled back to the at-rest position, a vacuum is created in the pump cavity so quick refilling of the pump is ensured. As pump pressure is relieved, the discharge check needle or ball reseats so there will be a closed check valve to refill the pump and so the signal created by air passing by the pump shooters will not pull fuel out of the pump system. The weight of this valve is designed to keep it closed against this signal. In some carburetors, the discharge check is a lightweight ball at the bottom of the pump passage. In this instance, fuel is maintained (stored) in the passage between the check and the nozzle (shooter). An anti-pull-over discharge nozzle is used in such systems so air passing by the shooter will not pull fuel out of the pump passage. Because a lightweight ball is used as the discharge check, excess vapors in the pump can escape easily into the passage to the shooter. This can be very helpful when a diaphragm-type pump is equipped with a rubber-type inlet valve because these valves seal so vapors cannot escape back to the fuel bowl. Vapors can only be purged from the pump when the pump is operated or when pressure becomes sufficient to raise the discharge ball or needle off of its seat.

Clearance around the hanging-ball-type inlet check in a diaphragm pump or around the pump stem in a piston-type pump allows excess vapors to escape to the fuel bowl.

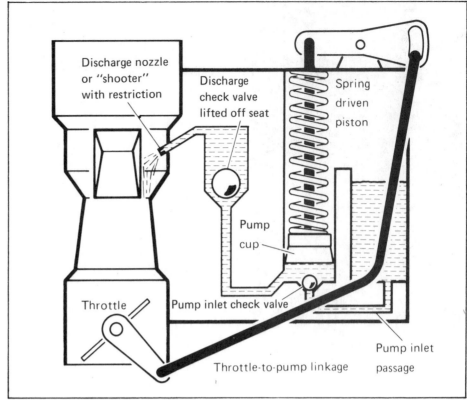

Here the throttle has been partially opened, causing the accelerator-pump cup to move to the bottom of its well. The pump-inlet check ball (valve) is forced onto its seat and the discharge check ball (valve) has been lifted off of its seat. Fuel is being sprayed into the carburetor through the "shooter." In many carburetors of similar construction, the pump cup serves as the inlet valve when the pump moves upward to refill.

Comparison of two pump-inlet-valve types. Rubber inlet valve at bottom is a rubber umbrella-shaped valve. It seats instantly for better tip-in performance than provided by hanging-ball type above. Hanging-ball type requires an instant seat before pump shot can be delivered. And, a tiny part of the pump shot is wasted into the bowl as pump pressure causes the ball to seat. A clearance of 0.011 to 0.013-inch is needed between the ball and its retainer (arrow) when the bowl is inverted.

33

Spring-driven accelerator pumps used on some Holley carburetors fill through center of pump cap. Cup has been cut away to show how it fits loosely on stem to allow filling when closing throttle linkage "cocks" pump against spring. Cup seals against plastic pump piston face during delivery stroke.

DESIGN FEATURES/Accelerator Pump System

Accelerator Pump Inlet Valves—The synthetic-rubber-type valve takes the least pressure difference to fill. Because it is normally closed, no time delay or fuel loss is required to close it and it has the best ability to supply a pump shot with very little throttle movement. This is called *tip-in* capability. It will provide an immediate pump shot as the throttle is opened quickly. A rubber valve is not as good as the ball-check valve for elimination of vapors, and it cannot be used with exotic fuels.

A ball-type valve is often used in accelerator pumps. It must be forced against its seat by fuel escaping past it. Thus, the ball type is not as good for tip-in as the rubber-type valve because some of the pump shot is lost. This is due to the delay required for seating the ball. The ball-type valve is especially good for purging vapors which may be generated in the quantity of fuel pocketed in the pump. Vapors easily escape through the clearance area between the ball and the seat. Exotic fuels do not affect the steel ball.

Two Pump Types—Holley normal replacement carburetors use both piston and diaphragm-type accelerator pumps.

All Holley high-performance carburetors use diaphragm-type accelerator pumps because these provide the most positive action. These pumps are available with maximum capacities of 30 cubic centimeters and 50 cc with the capacity selected to fit application requirements. Accelerator-pump capacity is measured by collecting the output of 10 full strokes of the accelerator pump. Thus, a 30 cc pump delivers 3 cc per pump shot at maximum stroke. Pump capacity and delivery rate are controlled by the pump cam or linkage settings.

Pump Cams—The high-performance carburetors use a nylon cam on the throttle lever to operate the pump lever. A "white" cam is typically supplied on the carburetor. The shape of this cam affects the pump as follows. First, total lift of the cam affects the stroke and therefore the capacity available from the pump. Cam profile or shape controls the phasing of the pump system.

A sharp-nose pump cam gives a quicker pressure rise and causes strong pump action to begin immediately as the pump is activated. A gentle-ramp cam gives the opposite effect.

The pump cam is used to make the carburetor perform well in a particular application. It is one more tool in the tuner's kit. Several cams which are offered were developed to specific applications. All pumps have adjustment features on either the cam or the pump linkage.

Pump Discharge Nozzles (shooters)—The accelerator pump discharges through a shooter. The shooter's hole size governs the rate of discharge. A larger hole allows the pump contents to be discharged quicker and with less pressure than a small discharge hole.

Discharge hole size in thousandths is stamped on removable shooters. A 25 marking indicates 0.025-inch holes. Typical sizes are 0.021, 0.025, 0.028, 0.031, 0.035 and 0.037 inch.

Removable pump "shooters" or discharge nozzles are used in all Holley high-performance carburetors. Several types of shooters are used and they are not interchangeable between carburetor types.

Typically, the pump shooter is aimed or targeted so fuel from the pump "shot" hits the booster venturi (if used). This "breaks up" the fuel so that it is more nearly vaporized as it enters the engine. Some may be targeted toward the throttle plate or against the bore.

It is well to remember that the function of the pump is to replace fuel which has dropped out of suspension and to compensate for fuel inertia or delay.

Pump Override Spring—The pump override spring is an important design feature of the accelerator pump, both in piston and diaphragm-type pumps. This spring controls fuel pressure in the accelerator-pump system and pump shot duration during wide-open-throttle "punches" or "slams." When the cam or link has lifted or pushed the pump linkage to its maximum lift point, the override spring takes up the full lift travel and then applies pressure to operate the piston or diaphragm. Without the override spring, something would have to give or break because fuel is not compressible.

Pump Selection and Timing—More details on pumps, cams, shooters, etc., are in the performance tuning chapter.

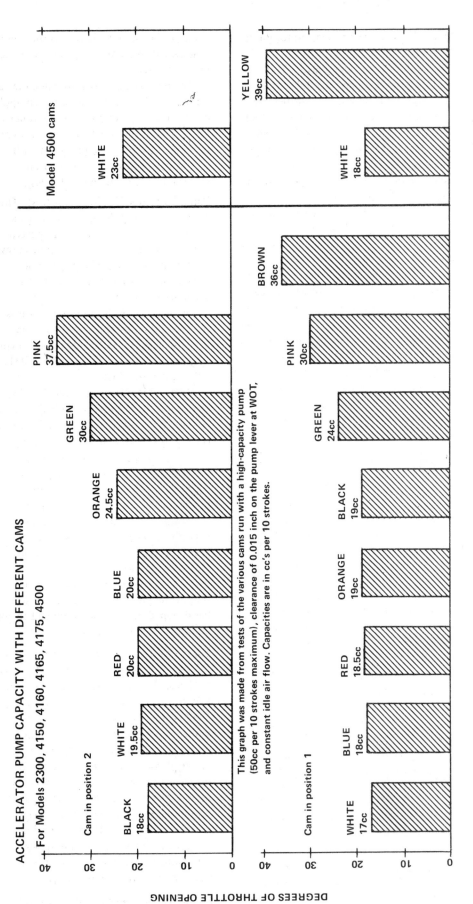

ACCELERATOR PUMP CAPACITY WITH DIFFERENT CAMS

For Models 2300, 4150, 4160, 4165, 4175, 4500

Model 4500 cams

Cam in position 2

	Throttle opening
BLACK 18cc	
WHITE 19.5cc	
RED 20cc	
BLUE 20cc	
ORANGE 24.5cc	
GREEN 30cc	
PINK 37.5cc	
WHITE 23cc	
YELLOW 39cc	

Cam in position 1

	Throttle opening
WHITE 17cc	
BLUE 18cc	
RED 18.5cc	
ORANGE 19cc	
BLACK 19cc	
GREEN 24cc	
PINK 30cc	
WHITE 18cc	
BROWN 36cc	

This graph was made from tests of the various cams run with a high-capacity pump (50cc per 10 strokes maximum), clearance of 0.015 inch on the pump lever at WOT, and constant idle air flow. Capacities are in cc's per 10 strokes.

DEGREES OF THROTTLE OPENING

Accelerator-pump discharge nozzles are targeted so the pump shot "breaks" on the booster venturi. Photo taken with carburetor on Holley air box shows how shot is pulled toward lower edge of booster venturi by air streaming into the carburetor. Bubbly fuel flow from main discharge nozzle issuing from tail of booster indicates air is already mixing with the fuel.

Typical routing of pump passages are used in carburetors with removable fuel bowls and metering bodies.

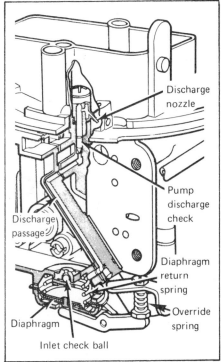

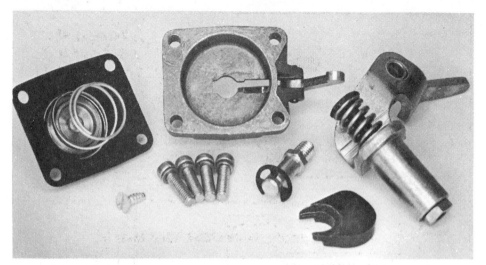

High-capacity (50cc per 10 strokes) accelerator-pump kit 85BP-3354 is often added when large secondaries are being opened quickly or where the engine has a wild camshaft. They are also helpful when the carburetor is a long way from the intake ports. The pump is often called a "Reo" pump because similar ones were used on Reo trucks. Installation may require raising carburetor with a 1/4-inch-thick aluminum spacer so pump lever will clear the manifold.

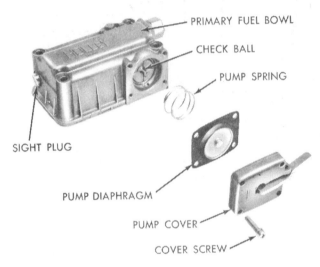

Diaphragm-type pump in removable fuel bowl showing component parts. Check ball is used as inlet valve in this example—some use rubber-type inlet valves. Pump is actuated by lever which rides against a cam on the throttle shaft.

Holley's pump-discharge-nozzle screw kit 85BP-5229 includes this screw which allows increased fuel flow to the discharge nozzle. When using a shooter size larger than 0.040 inch, one of these screws should be used as it ensures the nozzle will be the limiting restriction. This is for Models 2300, 4150/4160, 4165/4175 and 4500.

Accelerator-pump discharge nozzles or "shooters" are marked with numbers indicating diameter of holes in thousandths of an inch. Three types shown here include anti-pullover design used in 4165 and other models with lightweight pump-discharge-check valves. Two shooters at right are used with heavier check valves located immediately under the shooter.

Choke System

The choke system provides the richer mixture required to start and operate a cold engine—discussed in the fuel requirements section. Cranking speeds for a cold engine are often around 50 to 75 RPM. These speeds are low compared to engine operating speeds, hence very little manifold vacuum is created to operate the idle system. A closed choke valve creates a vacuum below it so fuel is pulled out of both the idle and the main-metering systems during cranking.

As you might imagine, this surplus fuel creates an extremely rich mixture of approximately equal fuel and air. The super-rich mixture is needed because there is not much manifold vacuum to help vaporize the fuel. Because the manifold is cold, most of the fuel puddles onto the manifold surfaces as it immediately recondenses. Also, the fuel is cold and not volatile. Not only that, but the fuel which issued from the main metering system was largely liquid because there was no air velocity to assist in atomizing it. Liquid fuel cannot be evenly distributed to the cylinders and when it arrives there, it will not burn correctly. So, only a small portion of the fuel ever reaches the cylinders as vapor during starting.

Once the engine starts, off-center mounting of the choke plate shaft allows air flow to open the choke slightly against a spring to start leaning out the mixture.

In the case of the automatic choke, a vacuum "qualifying" diaphragm pulls the choke valve to a pre-set opening once the engine starts. In some cases a temperature-modulated diaphragm varies this pre-set opening with ambient temperature. When the choke assumes the qualifying position it is still providing a 20 to 50% richer than normal operating mixture as the engine warms up and the choke "comes off" or opens.

This richer mixture is required because fuel will not vaporize correctly until the exhaust hot spot in the manifold has warmed up sufficiently to ensure good vaporization. So long as the choke is "on," the engine idles quite fast—usually 800 to 1100 RPM or higher with a cold engine—because the choke linkage also includes a fast-idle cam to keep the engine running

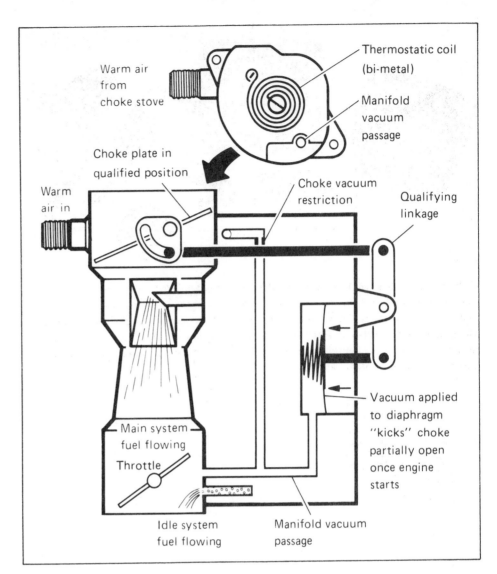

Warm air from choke stove

Thermostatic coil (bi-metal)

Choke plate in qualified position

Manifold vacuum passage

Warm air in

Choke vacuum restriction

Qualifying linkage

Main system fuel flowing

Throttle

Vacuum applied to diaphragm "kicks" choke partially open once engine starts

Idle system fuel flowing

Manifold vacuum passage

Choke plate shape—The choke-plate closely fits the air-inlet-horn contour so that a vacuum will be created below it when it is closed. Some choke plates are slightly bent. This may relate to the qualification setting or to mixture quality. The need for such configuring of the plate was established by the carburetor design engineers during development.

fast to aid in vaporizing fuel and in over-coming cold-engine friction loads.

As the engine warms up, the choke is opened fully by the bi-metal spring in an automatic choke, or by the operator if the choke is a manual one. This action also reduces the idle speed to a normal "curb idle."

DESIGN FEATURES/Choke System

Automatic Chokes—There are two main types of automatic chokes: Integral and divorced. The *integral* type uses a tube to connect heated air from the exhaust manifold or exhaust-heat crossover to the bi-metal spring inside a housing on the carburetor. Some integral chokes use engine coolant to supply heat to the bi-metal for choke operation. The *divorced* type uses a bi-metal spring mounted directly on the intake manifold or in a pocket in the exhaust-heat passage of the intake manifold. A mechanical linkage from the bi-metal spring operates the choke lever on the carburetor.

Divorced chokes provide the most accurate reflection of engine requirements because the choke is only operated when the engine gets cold. An integral choke may be closed because there is no flow of heated air past the bi-metal spring when the engine is not running. So, the integral choke can close while the engine is still hot, even though a choke-supplied rich mixture and fast idle are not needed to start the engine and keep it running.

Electric Choke—Chokes can be operated by a bi-metal spring heated by an electric resistor providing increasing heat similar to that provided by the engine as it warms up. The choke thus operates as if the bi-metal spring were sensing temperature supplied by the engine. Advantages of an electric choke are simple hook up without plumbing and other attachments to the engine. Only a single wire is needed, usually from the ignition switch. Disadvantages of an electric choke are:

1. Current is taken from the battery when power demands are high.

2. There is a quick come-back-on which reinserts the choke whenever the engine is turned off, even if the engine stays warm.

3. Unless the engine is immediately started when the ignition is turned on, the choke comes open, even though the engine is still cold. This problem can be fixed by powering the choke from the alternator circuit so the choke only gets power when the engine actually starts. Current can also be supplied through a pressure switch so the choke only heats when the engine is operating. Switch 89R-641A can be used for this with wires connected to the C and NO terminals.

4. The choke can come off while the engine is still cold.

Choke Index—Automatic chokes have index marks. The factory setting causes the choke to close (on a new carburetor) when the choke bi-metal is at 70°F. Tapping the carburetor lightly will overcome any shaft friction so the choke will seek the position being set by the bi-metal. If less choke is desired at this temperature, then move the choke index one mark. An arrow on the housing shows the direction the index must be moved to change the choke operating characteristic. A choke mixture change rarely requires more than one index mark change from the factory setting.

Unloader—If the engine does not start quickly, additional fuel which the choke has fed to the engine will have created an overrich mixture which cannot be burned. This must be cleared (or purged) from the manifold with air, which requires cranking the engine with the throttle held wide open, and opening the choke so additional fuel is not drawn in. On carburebors with automatic chokes, opening the throttle wide causes a tang on the throttle lever to contact the fast idle cam. This unloads (opens) the choke plate sufficiently to allow clearing excess fuel from the manifold.

Divorced choke is actuated by a remotely mounted bi-metal (arrow). It is often placed in an exhaust-heated pocket in the intake manifold. Divorced chokes are subject to greater production tolerances than integral types.

What's a bi-metal? The bi-metal referred to in the accompanying automatic-choke discussion is two different metals bonded into a strip and formed into a coil. Because the metals have unequal thermal-expansion characteristics, this coil unwraps when warmed and wraps up again when cooled. The outer or "free" end of the coil attached to the choke linkage holds the choke plate closed — or loads it to close the plate when the throttle is opened — until the bi-metal is warmed. Warming can be by exhaust-gas-warmed air, jacket water or an electric heating element — depending on the type of choke actuation. Bi-metal temperature-response characteristics are built-in according to the metals selected to make the strip. Most automotive-choke bi-metals wrap up (choke just closed) at 65° to 70° F.

Electric choke has nichrome-wire heating element to warm bi-metal and open choke. Power is supplied from a 12-volt source.

Examples, left to right: manual choke (arrows indicate cable-housing clamp and lever clamp for cable), integral choke with bi-metal and housing removed to show built-in vacuum-qualifier piston (arrow), divorced choke with vacuum-qualifier diaphragm (arrow) attached to carburetor exterior.

Vacuum-qualification setting for choke is detailed in repair-kit instructions and in the tune-up specifications provided in manual for car on which carburetor was supplied. In some cases the check is made on the down side of the choke plate — as here — in others on the up side.

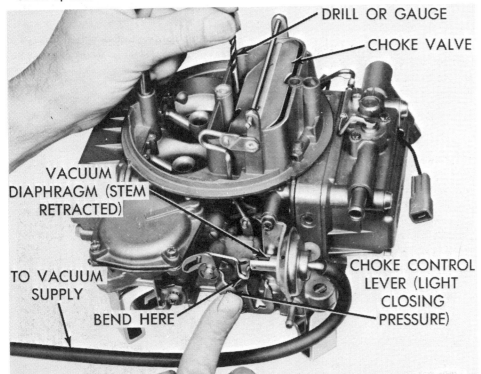

DRILL OR GAUGE

CHOKE VALVE

VACUUM DIAPHRAGM (STEM RETRACTED)

TO VACUUM SUPPLY

BEND HERE

CHOKE CONTROL LEVER (LIGHT CLOSING PRESSURE)

Choke index can serve as a guide when making changes to choke settings. Such changes should usually be limited to one mark RICH or LEAN as noted on the choke housing. Three screws attaching housing must be loosened to allow adjustment to be made.

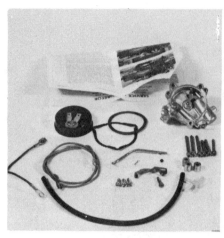

Electric choke conversion kit 85R-5175 uses choke vacuum through a rubber hose. Fits R-4412 two-barrel, R-6299, R-6708 and 6709 four-barrels, and "47" series double-pumpers (-2 or higher) such as R-4776-2, R-4777-2, etc.

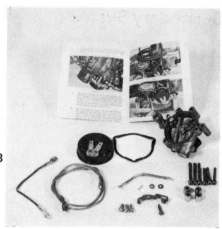

Electric choke conversion kit 85R-5178 converts carburetors with an internal choke vacuum supply. R-1850 and R-3310 four-barrels and the R-6425 and R-7448 two-barrels use this kit.

Electric heat Calrod next to bi-metal provides one more "handle" in programing choke come-off time to meet emission standards.

Adjustable unit allows changing bi-metal preload. Note the index marks on the adjacent stampings. 1972 and later model years cannot use this type because it can affect exhaust emissions.

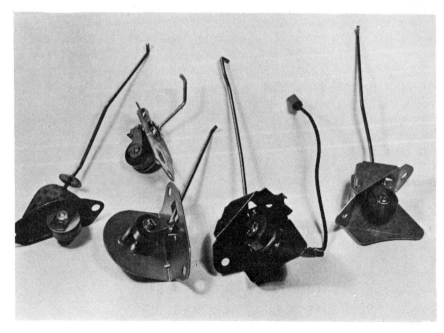

A collection of remote or divorced choke units found on GM and Chrysler vehicles.

Secondary System

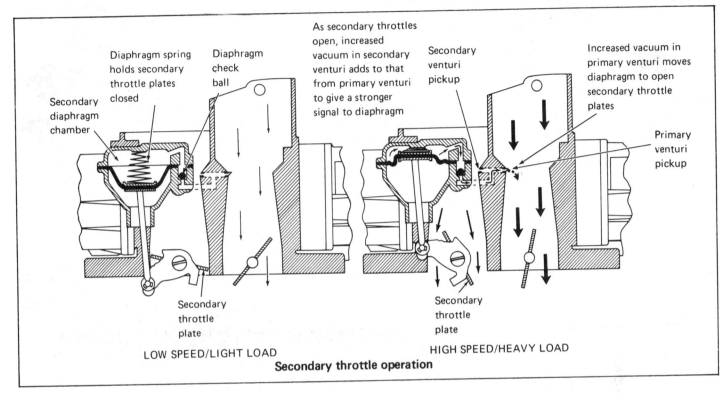

As secondary throttles open, increased vacuum in secondary venturi adds to that from primary venturi to give a stronger signal to diaphragm

Secondary diaphragm chamber

Diaphragm spring holds secondary throttle plates closed

Diaphragm check ball

Secondary venturi pickup

Increased vacuum in primary venturi moves diaphragm to open secondary throttle plates

Primary venturi pickup

Secondary throttle plate

Secondary throttle plate

LOW SPEED/LIGHT LOAD

HIGH SPEED/HEAVY LOAD

Secondary throttle operation

Schematic of diaphragm-operated secondary throttle. Signal from primary venturi is augmented by an increasing signal from secondary venturi as secondary throttle starts opening. Closing primary throttle closes secondary by mechanical override linkage. And, as signal from the venturi ports "dies" (pressure closer to atmospheric) check ball blows off seat to remove vacuum from diaphragm chamber. Opening rate of secondary is controlled by spring behind diaphragm and size of restriction through which vacuum is applied.

For many years U.S. carburetors were single-staged carburetors. The V-8 was equipped with a two-barrel, but this was nothing more than a single casting containing two one-barrel carburetors side by side—one for each level of the traditional two-level cross-H manifold. So consider the usual two-barrel as a single-stage carburetor.

The late 1940's saw an increased emphasis on vehicle performance. Because more air flow means more power, single-stage carburetors became bigger and bigger—and so did drivability problems. The larger venturis which were used required higher RPM or greater air flows to start main-system flow. Under some conditions, the idle system could be made to cover up the late entry of the main system. But, because the idle system is controlled by manifold vacuum, a great deficiency was felt at low manifold vacuum and low air flows, such as in a high-load, low-RPM condition. In addition, the low venturi velocities at low RPM resulted in poor vaporization of the fuel and poor fuel/air mixing. These caused cylinder-to-cylinder distribution problems and erratic operation of the engines at lower air flows.

Simply stated, the metering range of the single- or two-barrel carburetor was too narrow to satisfy all requirements. The answer was obvious; use a staged carburetor to stretch the metering range and get back to venturis small enough to get the main systems flowing at low RPM and provide good vaporization—while having the required capacity for high-RPM operation when it was needed.

Six-cylinder engines of medium displacement remained the economy workhorses, so a staged two-barrel was never introduced on a U.S. car until the 1970 Pinto, even though such systems had long been used on European and Japanese cars. There the costs were outweighed by the need to get maximum economy and maximum performance from small displacement engines.

However, in the early 1950's, a whole host of four-barrel carburetors were introduced for V-8's. The primary side of these carburetors was just like the older single-stage carburetors. Venturi size in the primary was smaller than it had been on the single-stagers, returning all of the benefits provided by small venturis: Early start of the main system, good vaporization and fuel/air mixing—and good dis-

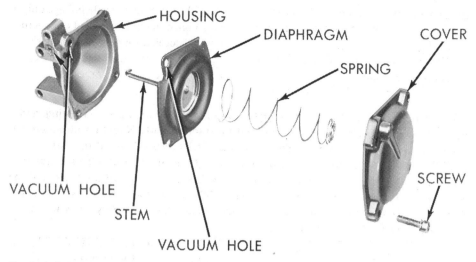
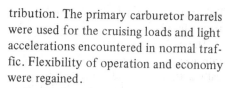

HOUSING

DIAPHRAGM

COVER

SPRING

SCREW

VACUUM HOLE

STEM

VACUUM HOLE

Exploded view of diaphragm used to operate secondary throttles. Spring is only element which should be varied when tuning these carburetors for a different opening point for the secondary throttles.

Diaphragm-operated secondary opens as the engine needs added air flow. This type of secondary operation is very "forgiving" in terms of size selection. A too-large carburetor (in air flow capacity) can often be used without spoiling low-end performance and drivability. Arrow indicates secondary-throttle stop.

Mechanically actuated secondary throttles operate progressively as shown here. Primary opens 40° or so before secondary opening begins. Linkage opens secondaries fully as primary throttles reach full-open.

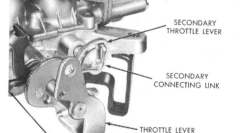

SECONDARY THROTTLE LEVER

SECONDARY CONNECTING LINK

THROTTLE LEVER

ACCELERATING PUMP LEVER

Primary throttle rod in slotted lever on secondary shaft will not allow secondary-throttle opening beyond an equivalent primary-throttle opening. Even though primaries may be opened wide, the secondaries will not open until there is sufficient air flow through the primary venturis to cause the diaphragm to actuate the secondary throttles.

tribution. The primary carburetor barrels were used for the cruising loads and light accelerations encountered in normal traffic. Flexibility of operation and economy were regained.

Coupled to the primary carburetor barrels were two secondary carburetor barrels—designed to operate when maximum air flow was required for more power. Essentially, the metering range of the carburetor was doubled, combining good part-throttle operation with relatively unrestricted flow for maximum-power conditions.

The secondary side of a carburetor is simply another carburetor which usually opens later than the first. Secondaries always have their own main-metering system. Most have an idle system, too. This gives better mixture distribution and idle stability. The result is improved emissions performance because leaner idle settings can be used. The idle system also keeps the fuel level from rising in the secondary bowl when the needle and seat leak slightly or are opened due to float bouncing. A further advantage of the secondary idle system is it ensures that fuel in the secondary bowl is fresh.

In 1967 Holley began providing some carburetors with accelerator pumps for both the primary *and* secondary barrels. Some of the carburetors have power systems in the secondary side.

Operation of the secondaries can be accomplished in several ways:

1. Mechanical only
2. Diaphragm
3. Mechanical with velocity valve
4. Mechanical with air valve

Holley carburetors use the first two of these principles.

DESCRIPTION OF OPERATION

Mechanically operated secondaries are very simple in operation. Secondary throttles are opened by a direct link from the primaries, usually on a progressive basis so the secondary opening is delayed until the primaries have reached approximately 40° of opening. Closing the secondary throttles is very positive because of the return spring, because the throttles are offset on their shaft so that air flow aids their closing and because a return link from the primary throttle pulls them closed.

Holley has pioneered the use of separate accelerator pumps on the mechanically actuated secondaries and this is now considered to be an essential feature for very high-performance applications. The secondary pump ensures that adequate fuel will be available to carry the engine through any period when the driver opens the throttles quickly. This is especially important when they are opened at an RPM too low to provide sufficient signal for main system flow on both the primary and secondary sides of the carburetor. The pumps actually provide mechanical fuel injection during the time the signal is being reestablished.

Diaphragm-operated secondaries have also been highly developed by Holley. These carburetors use a diaphragm to open the secondary throttles. Vacuum for the diaphragm is obtained from one of the primary venturis and part of this signal is bled off through an opening or bleed into one of the secondary venturis. As the engine is operated at increasingly higher RPM, velocity through the primary venturi creates a vacuum signal. The amount of secondary opening depends initially on air flow through the primary barrel, but as the secondary throttles begin to open and flow is established in the secondary barrel, the vacuum signal from the primary is augmented by vacuum from one of the secondary barrels through the bleed opening which was previously mentioned.

At maximum speed the secondary throttles are opened fully. If the carburetor is too large for the engine, the diaphragm automatically sizes the carburetor so that it flows only the needed amount of mixture by *partially* opening the secondary throttles. The secondaries remain closed when the primary throttles are opened wide at low RPM. This eliminates tiny bogs and allows use of the carburetor with a wide range of engine displacements, gear ratios, car weights, etc.

A ball check in the vacuum passage to the diaphragm applies vacuum at a controlled rate through a bleed (a groove in the check-ball seat). When RPM is reduced so the secondary air flow is not needed, the ball check instantly releases vacuum from the diaphragm so the throttles are closed by the return spring behind the dia-

phragm and by the action of the air flow against throttles which are offset on their shaft.

The diaphragm-operated secondary has an especially low opening effort or pedal "feel" because the throttle linkage only has to open the primaries and only actuates one accelerator pump. Secondaries are opened by the diaphragm without any assistance from the driver.

Chart and table below clearly indicate relationship of secondary diaphragm springs and opening points for each with two popular engine displacements. Note how displacement affects opening point. A larger engine opens secondaries earlier. The bottom spring (white) is a clear or "plain" steel spring. The slope of the lines indicates the secondary opening rate. A shallow slope opens more quickly. The springs toward the bottom of the graph allow sooner secondary opening. A yellow 38R-585 spring is the fastest-opening spring supplied in the kit 85BP-3185. The black spring delays secondary opening the longest and opens them at the slowest rate.

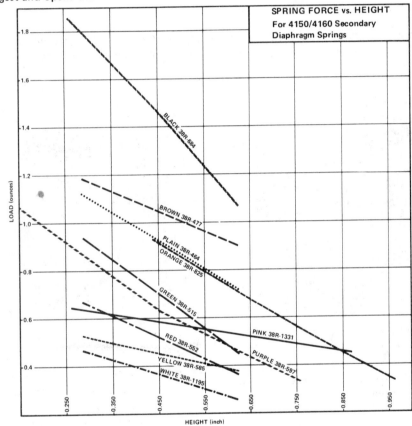

SECONDARY THROTTLE OPERATION RANGES
Diaphragm Secondary Springs From 85BP-3185 Used
in Model 4150, List 3310-1 Carburetor

Spring Color	350 CID Engine		402 CID Engine	
	RPM to Open	RPM at Full Open	RPM to Open	RPM at Full Open
Yellow (short spring)	1620	5680	1410	4960
Yellow	1635	5750	1420	5020
Purple	1915	6950	1680	6050
Plain (Std. Spring)	2240	8160	1960	7130
Brown	2710	8750	2380	7650
Black	2720	Not fully open at maximum air flow	2390	Not fully open at maximum air flow

NOTE:
All data taken without air cleaner. An air cleaner would cause earlier opening in all cases. Values subject to change due to cleaner restrictions.

Formula
$$CFM = \frac{RPM \times CID}{2 \times 1728} \times \eta$$

Where η = Volumetric Eff. = .9

Holley Tests
November, 1971

High-Performance & Replacement Carburetors

Bill Jenkins' highly successful 1974 332 CID Pro-Stock Vega. Bill likes to build his own induction systems. Main bodies are from List 4224 center-pumpers. Throttle bodies featuring 1-3/4 bores, are from List 4296. Both fuel bowls have rear inlets which makes two of them custom. Floats are hung from the rear. Bowl vents are a Jenkins' creation and the intake manifold is highly customized. Air horns are machined for hood clearance. Bill developed this package through thousands of hours of hard work. That's why he's a winner.
Photographs courtesy Jenkins Competition.

NOTE: Some Holley high-performance and replacement carburetors are designed to provide original-equipment emission performance and can be used for specific street/highway applications. Consult the current Holley Performance Parts catalog for details. Other Holley carburetors are designed for off-road/competition use and should not be used in any street/highway vehicle where local emission-control legislation prohibits.

A WORD ABOUT HOLLEY NUMBERS

A complete Holley assembly—such as a carburetor—can be referred to by three different numbers: Model number, engineering part or list number, and sales number.

Model Number—describes the class, type and general features of products with that group. Model 4150 describes a basic four-barrel-carburetor configuration. Many different calibrations and individual features exist on various part/list numbers within the same model grouping.

Model numbers usually have some meaning. The first digit of the model number usually designates the number of barrels (throttle bores). Example: Model 2210 and 2300 both start with the digit 2—and both are two-barrels. The other digits indicate variations making one different from the other. 4150 and 4160 are similar four barrels with minor secondary-metering-system differences.

As with any numbering system, there are exceptions. For instance, models 1901 and 5200 are two-barrels. For some reason, these were assigned "wrong numbers." Be assured that your Holley dealers won't think these are wrong numbers when you want to buy them.

Engineering Part or List Number—assigned to carburetor or other assembly in strict numerical sequence without regard for any other factor. If a two-barrel carburetor, a fuel pump and a four-barrel carburetor were initiated in that succession, the part numbers will follow that order. These numbers are then assigned to their model groupings.

Engineering numbers were originally

set up with an R prefix for the assembly number and on group numbers of components. The prefix letter may be changed to assist identification, such as P for fuel pumps and D for distributors.

Although Engineering Part and List numbers are now used interchangeably, the list number was originally used as an assembly's calibration identification. Part numbers and list numbers are now identical. Example, carburetor R-1234-1AAA is list number 1234-1. The list number is stamped on the carburetor, usually on the air horn extension supporting the choke shaft.

Sales Number—an identifying number assigned in numerical sequence for popular replacement carburetors and service items such as needle/seat, pump piston, pump diaphragm, repair kits, gasket kits, etc.

Sales numbers include a prefix number, a dash and a suffix number. Prefix 1 identifies carburetors, with the suffix identifying an individual carburetor. 1-108 completely identifies a particular carburetor. The number is used for merchandising and computer data processing and does not appear on the carburetor, even though it may be used on the carton and in price sheets.

Carburetor-service components are also identified by product categories or groups. Prefix numbers are assigned to screws, gaskets, throttle plates, etc. Suffix numbers identify the particular parts.

WARNING! Model 2300 throttle bores ARE NOT CENTERED. They are offset toward the rear of the carburetor base. If you are careless when installing the gasket, it can prevent correct throttle operation. The engine may start and run, but the throttles may snag in the gasket and not close.

Part No. 4412 Model 2300 has center-pivot race bowl with adjustable needle/seat and sight plug for setting fuel level without disassembling carburetor. High-capacity accelerator pump (50cc per 10 strokes) is used. Hand-choke linkage is standard on this one. A universal throttle linkage adapts the carburetor to Ford, GM and Chrysler products. This carburetor is designed for use on modified engines. It is not suitable for street use on stock 289 or 302 CID engines.

Adjustment and repair of these carburetors is described in the section starting on page 165.

Model 2300

Holley Model 2300 carburetors are most familiar to performance enthusiasts as those units used on three-carburetor Chevrolets and Dodge/Plymouth "Six-Packs." And many racers who compete in drag classes allowing only single two-barrels have become old buddies with the List No. 4412 500 CFM version of the 2300. Readers who are tune-up men or mechanics will recognize these as the reliable carburetors used for many years on Ford passenger cars and trucks, Chevrolet trucks, International Harvester vehicles, American Motors vehicles, White Motors, Reo, and Willys Jeep.

The Model 2300 has been supplied with venturis ranging from 1-inch to 1-7/16-inch diameter and air flows from 210 to 600 CFM.

Some diaphragm-operated 2300 carburetors are used as the end units on triple manifolds. These have no chokes or accelerator pumps and use metering plates instead of metering blocks, as described in the 4160 information. In every instance, the center-mounted 2300 is equipped with a choke, accelerator pump, and a metering block with screw-in main jets and a power valve.

Because the 2300 is literally the front half of a 4150/4160 carburetor, operation is exactly as described for the Model 4150 a few pages on. The only exception is the 600 CFM List No. 6425 which uses an accelerator pump and fuel-bowl combination like that in the Model 4165.

Some 2300's have a "reverse" idle adjustment as described and illustrated in the 4150/4160 section. The adjustment screws in to richen the mixture and backs out to lean it—just the opposite of what has always been done in the past. A label is on the metering block of carburetors with this changed idle system.

Two Model 2300 carburetors which are described in the following paragraphs are of most interest to racers and mechanics who want the utmost performance from a two-barrel.

List No. 4412—The 500 CFM version of the Model 2300 was first introduced in 1969 and became an instant success among those engine builders and tuners

Diaphragm-operated outboard carburetors on 3 x 2 installations such as this Chrysler setup use vacuum signal from center carburetor to operate the outboard units as the engine can use the extra air flow. Only the center carburetor has an accelerator pump. Care is required in installation so no distortion of the bases will occur (through uneven tightening onto manifold). Such can bind the throttle shafts, causing erratic operation. All throttle shafts must work very freely. Manual linkage which operates all throttles simultaneously is not recommended. Holley offers 500 CFM outboard carburetors, Part No. 4783, which have accelerator pumps. These are designed for use with a staged mechanical linkage (progressive action) in conjunction with Part No. 4782, a 350 CFM unit used in the center. The 4782 has an automatic choke and an accelerator pump. Linkage should start outboard carburetor opening when center unit is opened 40°.

racing in classes which limited carburetion to a single two-barrel. The two 1-3/8-inch venturis are fed by No. 73 main jets and the power valve is a 5.0-inch unit. A high capacity (50 cc per 10 strokes) accelerator pump is used. The center-pivot-float race-type bowl has a sight plug and externally adjustable needle and seat assembly.

The carburetor base is equipped with hose connections for timed and vacuum spark advance (the timed port can also be used to purge charcoal canisters) and for a PCV valve. The SAE 1-1/2-inch spread-bore pattern (3-1/2-inch x 5-1/2-inch hole spacing) fits directly onto Ford two-barrel manifolds on engines from 289 through 390 CID. Using the carburetor on other engines requires an adapter to mate the bolt pattern and 1-11/16-inch throttle bores of the 4412 to the manifold flange. Adapter 85R-3584 is used to fit the carburetor onto SAE 1-1/2-inch 2V pattern used on some late Chevrolets. Chrysler products (318 CID) and some early Chevrolets have a SAE 1-1/4-inch 2V pattern which can be mated to the carburetor with Adapter 85R-3583.

The universal linkage works with all automatic and standard transmission throttle arrangements, including the Ford kickdown rod hookup.

A 5-inch diameter air-cleaner mounting base is provided by the carburetor body. **List No. 6425**—A two barrel with 600 CFM air flow! Let that sink into your mind as you imagine what that means to class records for boats and cars where two barrels are the required fuel-metering equipment. These carburetors flow 600 CFM as they come out of the box, but if you'll take five minutes to remove the hand choke and install a velocity-stack kit, you can up the air flow by 8.3% to 650 CFM.

Holley engineers had to use a lot of ingenuity to get an increase of 150 CFM as compared to the previous "big two-barrel." First, they eliminated the boost venturis and that required a novel way to discharge fuel into the air stream. As you can see in the accompanying photos, 11 holes in the circumference of each venturi serve as discharge ports or nozzles. An annulus (groove) behind the holes connects to the main metering system. The 1-7/16-inch venturis are machined as separate parts and pressed into the main carburetor body. Slimmed-down throttle shafts were the second device used by Holley in search of more air flow.

On a two-barrel carburetor with venturis of this size, off-idle tip-in performance becomes very critical. No driver likes to have an off-idle bog—and this is always a real possibility when a large carburetor is installed, especially if it is mounted on a small engine. Although the high-capacity accelerator pump can deliver up to 5 cc's per stroke, the really important thing is for the pump to discharge fuel at the slightest throttle movement. Because the usual ball-type inlet check valve requires just an instant to seat before pump delivery can begin, Holley chose the rubber-type "umbrella" inlet check valve like they use on the 4165. With this pump arrangement, the check valve seals instantly and the pump operates with only a very minute movement of the throttle lever. And, the pump fills instantly with only a slight pressure differential. The discharge ball check seats quickly to create a partial vacuum in the pump cavity to ensure quick filling as the throttle is moved back toward a closed position. Locating the check valve close to the bottom of the pump-discharge passage tends to keep the passage filled with liquid fuel. And, the lightweight ball tends to allow vapors to escape from the pump cavity when the carburetor body heats up when the engine is stopped or idling for a long period. The anti-pullover pump-discharge nozzle prevents fuel—which remains in the passage connected to the nozzle—from being pulled into the airstream.

Fuel bowl, metering block and gaskets are the same as used on Model 4165 and they cannot be replaced with seemingly similar parts for other 2300 or 4150/4160's. The main item to remember is that the pump discharge will be obstructed if the wrong type of gasket is used. These differences are fully explained in the 4165 section. With the exception of the discharge nozzles which ring the venturis, operation of the 6425 is identical to that explained for the Model 4165 in that section.

The center-pivot float bowl has a sight plug and externally adjustable needle and seat assembly.

Hose connections are provided for timed and vacuum spark (the timed port can be used for purging a charcoal cani-

Front view of List 6425 shows hand choke linkage and high-capacity accelerator pump. Accessory velocity stacks improve air flow capacity as described in the text. Stacks, Part 85R-4235, will not fit other Model 2300's without a lot of modification.

Slabbed or reduced-section throttle shaft is one of the methods Holley engineers used to improve 6425's air flow. Discharge-nozzle holes are visible around venturis.

Discharge holes around periphery of venturis are used instead of the more usual booster-venturi/discharge-nozzle combination in the 6425. Accelerator-pump discharge nozzle is anti-pullover design used with a ball-check valve in lower portion of metering block.

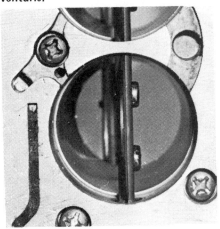

ster) and for a PCV valve.

As with the other 2300's, the carburetor has an SAE 1-1/2-inch spread-bore base which fits Ford two-barrel manifolds. Adapters can be used to mate the bolt pattern and 1-3/4-inch throttle bores to other engines as described in the preceding section on the List No. 4412 500 CFM unit.

The universal-type linkage works with all automatic and standard transmission throttle arrangements, including the Ford

kickdown rod hookup.

A 5-inch diameter air-cleaner mounting base is provided by the carburetor body. A 2-inch clearance should be maintained above the velocity stacks. Remember this when you install an air cleaner on this carburetor with stacks. This may require using two open-element cleaners glued together with Silastic RTV compound so there will be no chance of dirt sneaking in between the two elements.

List 6425 has a universal throttle linkage. Photo at left shows Ford installation (a spring and spring bracket are not shown here), center is Chrysler and Chevrolet is on the right.

Acknowledgments

Two authors' names are on this book, but it took far more than two to create it. The outstanding contributor to the first Holley book was Bob Miller, Engineering Section Supervisor for High-Performance Carburetion at Holley. He lived with the project, gathered much of the interesting information and provided enormous enthusiasm to help us finish the job. Another Holley engineer, Technical Services Director, Bob Lift pointed out additions and corrections needed to make everything crystal-clear.

Myron Lawyer, former Service & Education Manager at Holley, made all his drawings and photos available and Bill Krygier rounded up the helpful exploded views. Mark Campbell supplied a lot of the photos.

Andy Guria, Holley P.R. and Advertising Manager, provided continuous encouragement and photos.

We borrowed some from Doug Roe's H. P. Book *Rochester Carburetors*.

Automotive journalist, Roger Huntington, helped with quotes from his manifold-design articles.

Chevrolet Public Relations' Jim Williams and Dick Maxwell of Chrysler Corporation came up with materials for the high performance tuning sections. Ford's Chuck Mulcahey of Public Relations and Charles Gray of the engines group searched their files and provided photos we could not obtain anywhere else.

Dino Fry helped make sense of the complicated manifold-gasket situation for GM machinery.

Special thanks go to Mike's son Steve and daughter Sue who took turns supplying the necessary "third hand" during the repair photography sessions.

All of these fine people helped make the book more accurate, understandable, interesting and pictureful.

We started revising the first edition almost from the day of its birth. Comments and suggestions were received from many sources. We made changes wherever we felt the book could be bettered in any way. You, the reader, benefit because the book is finer and fuller than the two of us could have made it. To all of those who helped—including those whose names we have not mentioned—we say THANK YOU!

Mike Urich
Manager, Holley Aftermarket Engineering

Bill Fisher
Publisher of H. P. Books

Models 4150 & 4160

Current Holley four-barrel carburetors fall into three basic families: 4150/4160 (and 3150/3160 three-barrels), 4165/4175, and 4500 versions are separately described because of important constructional and operational differences.

Holley 4150 carburetors were first introduced on Ford 312 CID-engined automobiles in 1957. And, while we like to think of these units as strictly high-performance devices, they have also been used on trucks, too. Some of the important 4150 features include:

1. Center-pivot float bowls, first used on Ford and Chrysler NASCAR racing engines in 1964. These were a "snout-type" bowl which is no longer offered. The current style center-pivot fuel bowls were developed for Chevrolet applications.

2. Externally adjustable floats with sight plugs, first used on 1957 Ford Model 4150's.

3. Bowl vent plastic "whistles," first used in carburetors supplied for 1966 Fords.

4. Double accelerator pumps, first used on List No. 4296 4150's for 1967 Chevrolet big-block applications.

5. Mechanical secondary linkage, first used for Chrysler in 1959 (List No. 1970).

6. Center-mounted accelerator pump discharge nozzle, designed for plenum-type tunnel-ram manifolds and first used in 1967 (List No. 4224).

A very wide range of 4150-type carburetors is offered for aftermarket use. These range from 390 CFM to 850 CFM. Vacuum-secondary types are available in 390 CFM through 850 CFM in stair-step sizes to fit almost any requirement. Double-pumper types with mechanical secondary actuation are available in 50 CFM increments from 600 through 850 CFM.

4150 and 4160 carburetors are so similar that you might not be able to tell them apart if you did not know what to look for. They are absolutely alike in operation and—except for the secondary metering block—their construction is also identical. The 4150 secondary metering block has the same physical dimensions as

Model 4150 double-pumper List 6708 and 6709 are emission-oriented carburetors designed for good street performance on engines without EGR through 1972. Most emission hookups are included. Dual-inlet bowls have center-pivot floats. The metering blocks are the 4165 type with the improved pump system. Primary bores are 1-1/2 inch; secondaries are 1-3/4 inch. 650 and 750 CFM units are available.

Adjustment and repair of these carburetors is described in the section starting on page 165.

the primary metering block and it is sandwiched between the fuel bowl and the carburetor body—as is the primary metering block.

A secondary metering plate with built-in orifices is screw-attached to the body of 4160 carburetors. This metering plate is simpler in construction. It requires less drilling, no plugging or tapping, and has no screw-in main jets or power valve. With the single exception of the 4224 "center-squirter" 660 CFM units for tunnel-ram manifolds, all of the current 4160 carburetors have vacuum (diaphragm) operated secondaries—none have mechanical linkage or double-pump arrangements. Any of the three fuel bowls may be used, depending on the requirements of the manufacturer; side-hung, side-hung with external adjustment for level setting, or center-pivot with an inlet at each fuel bowl.

4160's have been used by American

Motors, Chevrolet, Chrysler and Ford as original-equipment carburetors. They are just as high-performance oriented as any of the 4150's and they can be converted to any of the 4150 features with appropriate kits which include a separate secondary metering block with gaskets, longer fuel-bowl screws and a longer fuel-transfer tube. This is Kit 85R-3548. Other items, such as bowl vents and high-capacity accelerator pump kits, can also be installed.

Why the 4160? Well, if you are making automobiles by the thousands and can buy Holley performance for a buck or so less per carburetor—the dollar savings mount up rather quickly. The 4160 is merely a low-cost 4150 made so car makers could afford to put on a Holley on some high-performance packages. Also, the 4160 allowed fitting two carburetors onto a 2 x 4 manifold where this might not otherwise be practical.

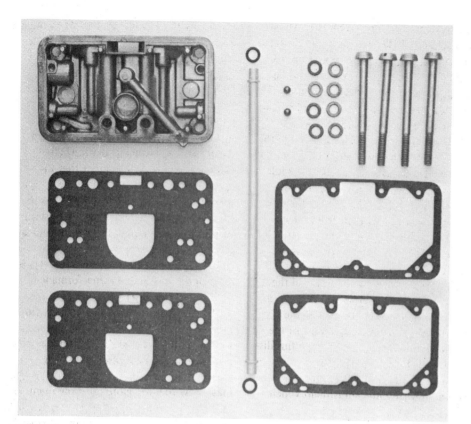

Kit 85R-3548 converts 4160's to 4150 configuration with a metering block containing replaceable main jets. Longer bowl screws, metering block, gaskets and a longer fuel-transfer tube are included. This kit is also used when converting a 4160 to use race-type center-hung fuel bowls. Note that the metering block does not have provisions for installing a power valve. Lead balls are included in the kit for closing the holes left when the balance tube is removed.

Secondary metering plate in 4160 is attached with clutch-head screws. Arrow indicates balance tube used in some of these carburetors to equalize fuel levels in the two bowls in the event of slight fuel leakage through an inlet valve (called "creep"). This is seldom a problem with Viton-tipped needles, but was often found with plain-steel needles which do not always seat perfectly.

DESCRIPTION OF OPERATION

This description is applicable to all 3150, 3160, 4150, 4160, 4165 and 4175, 4500 (except for intermediate system described in that section), and 2300 carburetors. No attempt has been made to explain why certain things happen in the functioning of the carburetor because these are fully covered in Chapter 2.

Fuel Inlet System—Fuel enters the primary fuel bowl through a screen or filter and passes through the inlet needle and seat assembly into the fuel bowl. Where there is only one fuel inlet for the carburetor, fuel reaches the secondary bowl needle and seat through a transfer tube O-ring-sealed to each fuel bowl. If there are two fuel inlets, fuel must be plumbed to each of these to ensure fuel for each side of the carburetor.

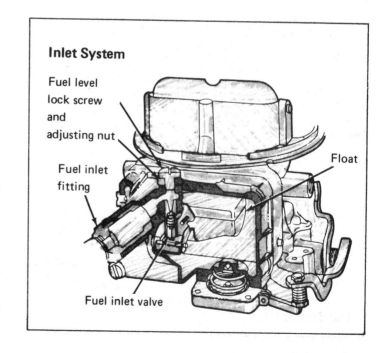

Inlet System

Fuel level lock screw and adjusting nut

Fuel inlet fitting

Fuel inlet valve

Float

Carburetors equipped with the "reverse" idle system have a label on the metering block to indicate that the adjustment is leaned by turning the adjustment screw counter-clockwise (admitting more air, as shown in drawing above).

Idle System—An idle system is provided for each barrel. The primary side is adjustable on the four-barrel units and on the **5200** and **5210** two-barrels. Other than the adjustment feature on the primary side, idle systems are essentially identical in construction and operation. At idle, fuel flows from the fuel bowl through the main jet and into the main well. From the main well, a passage connects to the idle well. Fuel flows through an idle-feed restriction into the idle well, up the idle well and is mixed with air from the air bleed as it flows down another vertical passage. At the bottom of this passage, idle fuel branches. One leg goes to the transfer slot above the throttle and the other goes past the adjustment needle to a hole in the throttle bore below the throttle plate. At curb idle, most of the fuel for idling is supplied through this discharge hole below the throttle.

Some **2300** and **4150/4160** carburetors have a "reverse" idle adjustment which screws in to richen the mixture and backs out to lean the mixture—just the opposite of what has always been done. Labels indicate the adjustment change. This difference came about as part of a development program to get improved idling with idle mixtures lean enough to pass emission requirements. The idle system was changed as shown in the accompanying drawing. Note the adjustable second air bleed in this new idle circuit. The mixture screw which once varied the mixture outlet area now varies inlet-air area through a second air bleed connected to the throttle bore just below the venturi. Holley engineers found the improvement in idling (they call it "idle quality") so great that the new idle system will eventually be incorporated in other carburetors. As this book went to press, this new idle system had been incorporated in the 4160 and 2300 aftermarket carburetors for late Ford automobiles. It is also in most 4165, 4175 and in the emission/performance 4150 double-pumpers.

The circuit "helps" some engines—but not others. For instance, it does not seem to help on a Chevrolet. The difference may be due to basic manifold design: The more usual single-idle-air-bleed system may allow fuel to "glob" into the manifold at idle—which some engines apparently tolerate without problems. Engines with manifolding not tolerant of such "globbing" probably idle better because adding the second air bleed to the idle circuit provides better idle-fuel vaporization—and improved distribution.

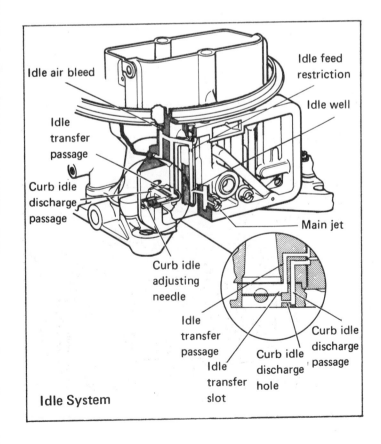

Idle System

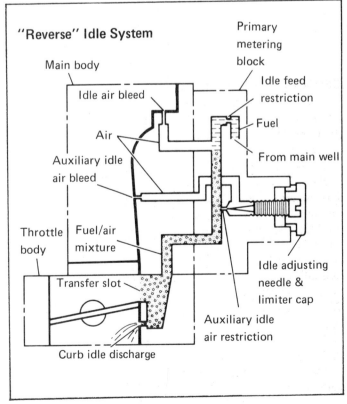

"Reverse" Idle System

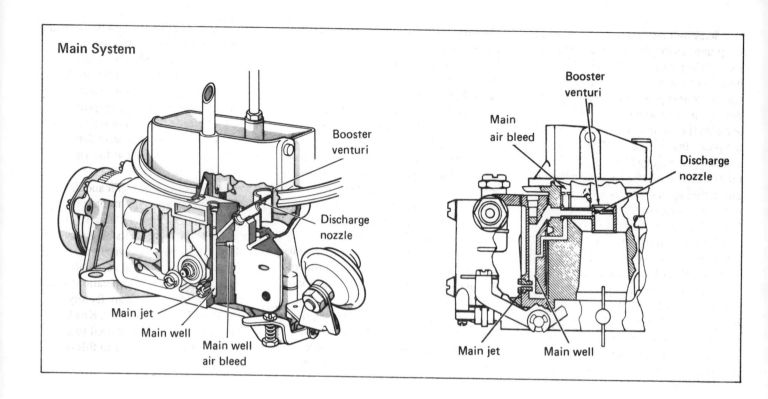

Main System

Booster venturi

Discharge nozzle

Main jet

Main well

Main well air bleed

Booster venturi

Main air bleed

Discharge nozzle

Main jet

Main well

Main Metering System—At cruising speeds, fuel flows from the fuel bowl, through the main jet into the bottom of the main well. Fuel moves up the well past air-bleed holes in the side of the well. These air-bleed holes are supplied with filtered air from the main air-bleed openings in the carburetor air inlet. The mixture of fuel and air moves up the main well and across to the discharge nozzle located in the booster venturi.

Power System—During high-speed operation or operation when manifold vacuum is low (near atmospheric), the carburetor provides added fuel for power operation. A vacuum passage in the throttle body transmits vacuum to the power-valve chamber in the main body. This manifold vacuum is applied to a diaphragm in the power valve to hold this valve closed at idle and normal load conditions. When manifold vacuum drops, a spring opens the power valve to admit extra fuel. This fuel flows through the power valve and into the main well where it joins the fuel flow in the main metering system to enrich the mixture.

When engine power demands are reduced, manifold vacuum increases so the power valve spring tension is overcome and the valve is closed, stopping the extra fuel flow.

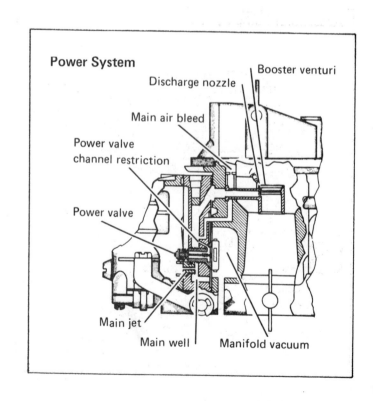

Power System

Discharge nozzle

Booster venturi

Main air bleed

Power valve channel restriction

Power valve

Main jet

Main well

Manifold vacuum

Accelerator-Pump System—The accelerator pump is a diaphragm type in the bottom of the primary fuel bowl (also in the bottom of the secondary fuel bowl in the case of double-pumper carburetors). The pump functions when the throttle linkage actuates the pump lever. Pressure forces the pump inlet check valve onto its seat and this prevents fuel from flowing back into the fuel bowl. Fuel flows from the pump through a long diagonal passage in the metering block to the main body. There the fuel pressure raises a discharge check needle off its seat so fuel can spray into the venturi from the discharge nozzle.

As the throttle is returned to a more closed position, the linkage returns towards its original position and the diaphragm spring forces the diaphragm down. The pump-inlet valve moves off its seat and the pump refills with fuel. As pressure is relieved from the pump passages, the pump-discharge check needle reseats so air flowing past the discharge nozzles will not pull fuel out of the pump passages and so the pump will refill.

Secondary System—Secondary systems are operated by a vacuum diaphragm or by a mechanical linkage connection to the primary throttles.

The secondary side of the carburetor has a fuel-inlet system, a non-adjustable idle system, a main metering system and—in the case of double-pumpers—an accelerator-pump system. These systems operate exactly like those on the primary side in nearly all instances. Some secondary systems have a power valve, some do not.

Mechanical Secondary—There are various types of mechanical-secondary linkage. Most use a direct link between the primary and secondary, with a slot in the secondary lever providing progressive action. Through 1975 a slotted primary lever with a roller/lever to transmit opening motion through a rod to the secondary throttles was used. This actuation is seen on the carburetor on the front cover. Opening the mechanical secondaries is controlled entirely by primary-throttle position. When progressive action is used, both primary and secondary throttles reach wide-open at the same time.

Choke System—Both integral and divorced choke operating mechanisms have been supplied on the various list numbers of the 4150/4160. Some units have manual choke mechanisms. A conversion kit is offered to convert integral automatic chokes to manual (hand) operation: 85R-3007. Some 4150/4160's are factory-equipped with electric chokes. Kits 85R-5175 and 85R-5178 can be used to convert 4150/4160 carburetors to this type of choke actuation.

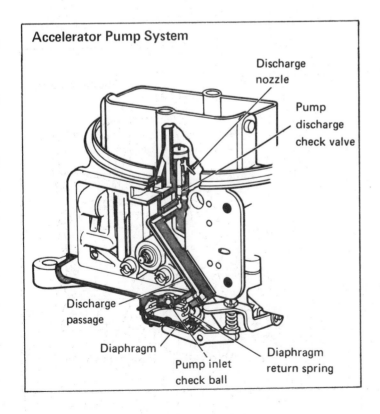

Accelerator Pump System

Discharge nozzle

Pump discharge check valve

Discharge passage

Diaphragm

Pump inlet check ball

Diaphragm return spring

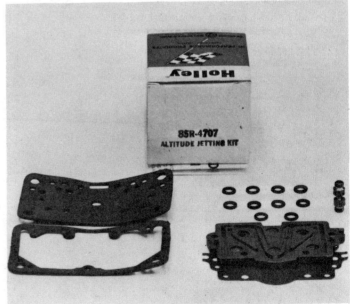

Altitude-jetting kit intended for RV series of carburetors used above 3000 ft. contains jets, gaskets and metering plate for List 6619, 6979, 6919, 6989, 6909 and 7009. **Photo by Mark Campbell.**

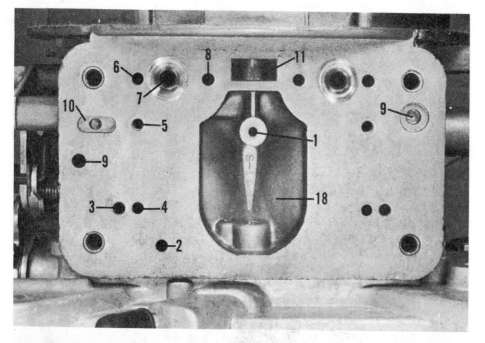

Description of holes in main body gasket surface joining to metering block:

1 — To discharge nozzle (accel. pump)
2 — To timed-spark port
3 — To curb-idle discharge
4 — To idle-transfer slot
5 — Used only with auxiliary idle air bleed
6 — To idle air bleed
7 — To main discharge nozzle
8 — To main air bleed
9 — Dowel locators for metering block
10 — Not used
11 — To bowl vent (pitot tube)

Description of holes and passages in metering block. Top photo shows side which mates to main body gasket. Lower photo is bowl side of metering block.

1 — Accelerator pump discharge passage
2 — Timed-spark passage (see 12)
3 — Curb-idle discharge
4 — Idle-transfer fuel connects to main body and to curb idle adjust screw
6 — Idle bleed air enters from main body
7 — Main passage to discharge nozzle
8 — Main bleed air enters from main body
9 — Dowels to position block & gasket
11 — Bowl vent passage
12 — Timed-spark tube boss
13 — Idle-mixture-adjustment needle
14 — Main jet
15 — Power valve threaded opening
16 — Power valve
17 — Power valve channel restriction (connects to main well 21)
18 — Manifold vacuum chamber (for power valve operation)
19 — Idle down well
20 — Idle well
21 — Main well
22 — Air bleed holes into main well
23 — Main air well
24 — Idle fuel from main well
25 — Idle feed restriction to idle well
26 — Fuel entry from accelerator pump in fuel bowl

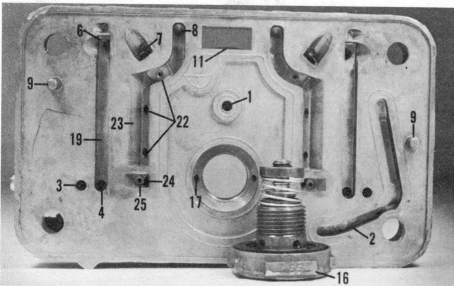

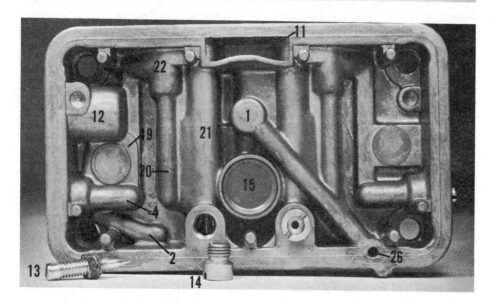

LIST 3310

One of the most popular carburetors in the entire Holley line is the List 3310 version of the Model 4150. This situation has been true since one version of the carburetor was introduced on the 425 HP 1966 Corvettes and 375 HP 1966 Chevelles—both with the 396 CID big-block Chevrolet engine. Other original-equipment uses of similar carburetors have included the famous Z-28 Camaro in both the 302 and 350 CID versions. Chevrolet versions of this carburetor have different list numbers.

Perhaps the reason for the popularity of the 3310 is due to its high-performance heritage and because it installs easily on a wide variety of cars with good assurance of working well on almost any 300 to 455 CID engine. Its flexibility of application comes from the diaphragm-operated secondaries which allow the carburetor to operate effectively over a wide range of air-flow requirements.

The carburetor has changed only slightly since its introduction. In 3310-1 form it was a 4150 type with two metering blocks. Its linkage and emission-connection features make it universally adaptable to Chevrolet, Chrysler and Ford V-8 engine applications.

As of 1976, the 3310-2 was introduced. This lowest-cost 780 CFM Holley performance four barrel has a 4160 metering-plate.

There are other diaphragm-secondary carburetors, but—with few exceptions—adapting these to cars for which they were not originally made can be a lot of work.

INSTALLATION NOTES

The 3310 carburetor has a universal throttle lever for Chevrolet, Chrysler, and Ford—including automatic-transmission models. The linkage parts shipped with the carburetor can be combined to create a throttle lever exactly right for your application.

One of the photographs on page 208 shows a 3310 plumbed for fuel inlet with Holley's fuel-line kit. Note the center-pivot float bowls. Each has its own inlet, so the fuel line must connect the stock fuel line to both inlets.

It is assumed, of course, that a single four-barrel intake manifold with the 5-3/16-inch x 5-5/8-inch bolt-hole pattern

List 3310 has vacuum operated secondaries, making it universally adaptable as a performance carburetor for 350—455 CID engines. Dual-inlet bowls and a hand choke are features.

is already installed on the car—or an adapter will be required. Adapters, unfornately, are usually "thieves." They steal air-flow capacity. As an example, a 780 CFM 3310 becomes a 640 CFM carburetor when an adapter is used between it and a manifold with the spread-bore pattern. It is always best to install a carburetor which fits the manifold directly. If the engine was originally equipped with a spread-bore carburetor (1-3/8-inch primaries and 2-1/4-inch secondaries), consider using a Holley 4165, 4175 or 4360 which will fit without an adapter. However, if one of these is not offered for your particular car, a 3310 with a replacement manifold would probably be a more logical choice because it has all of the right fittings and the correct throttle linkage for Chevrolet, Chrysler and Ford. Adapting a carburetor to a modern-day automobile is no easy task because there are so many connections and linkages to consider.

When you unpack the carburetor you will notice that all of the hose fittings are closed with rubber shipping plugs. These can be left installed when no hose will be attached to a fitting. However, if the 3/8-inch PCV fitting which extends from the base—just under the secondary bowl—at the rear of the carburetor is not used, replace this plug with a piece of hose plugged at one end. The plugged hose should be clamped onto the vacuum fitting so that it will not blow off in the event of a backfire—which is what the stock shipping plug will do if you attempt to use it. Or, if you prefer, the tube can be closed by filling it with an epoxy compound such as Devcon. Do not use rubber-type compounds to close such fittings (such as Silastic RTV) because these turn to jelly in the presence of gasoline.

All 3310-2 carburetors (current model as of 1976) have a manual choke. If you need a choke cable, buy a 6-foot-long hand control cable, 42R-466A. Electric choke kit 85R-5178 can be used if you prefer automatic choke operation.

Model 4160 RV carburetors are intended for use with emission/performance manifolds, particularly on recreational vehicles. Separate units are offered for General Motors, Ford, and Chrysler vehicles, with some made especially for vehicles with EGR (exhaust gas recirculation). All these carburetors have 1-9/16 primary and secondary bores, and flow 600 CFM. The secondary throttle is diaphragm-controlled.

This basic carburetor started out as a little brother to the 3310 and can be used as a good compromise between performance, drivability and economy on small-block engines. All emission hook-ups for recommended applications are included, so passing emission inspections should be no problem. These carburetors use the "reverse" idle system with limiter caps.

This gasket conversion kit allows using the cut-out type 8R-1572 RC or 8R-1685 RC metering block gaskets with carburetors not having accelerator-pump transfer tubes. The kit includes a roll pin, an O-ring (arrow) and a cut-out gasket. The 85R-5117 kit for Models 2300, 4150, 4160 using 8R-1023, 8R-1571 RC or 8R-1685 RC gaskets should be used with 0.040 inch or smaller pump shooters. Kit 85R-5118 fits 4165 carburetors using 8R-1023/1571/1684 gaskets.

O-ring-sealed accelerator-pump transfer tube is used on all Holley four-barrels since 1975. This tube provides a positively sealed passage for the accelerator-pump shot so none will be wasted. Note how gasket is cut to clear the tube. Any attempt to use the old-style non-tube metering block with the newer gasket will draw fuel from the accelerator-pump system and a full rich mixture will be supplied to the engine. Old-style metering block is shown at left in the lower photo.

Small-primary/large-secondary throttle-bore concept is generally conceded to be the plan for future carburetors, whether high-performance-oriented or not. Holley's 4165 series was the first spread-bore design combining high-performance capabilities with the ability to provide specification emission performance at the same time. Note the two accelerator pumps and mechanical secondary linkage. These carburetors are designed as bolt-on replacements for GM cars.

Model 4175. Diaphragm-operated secondary throttles make this carburetor ideal for spread-bore applications where heavy loads occur at low speeds, as with trucks and camper-type vehicles.

Models 4165 & 4175

Holley's Spread-Bore 4165 carburetors were introduced to the automotive world in 1971. These bolt-on replacements found instant favor among owners of cars equipped with other small-primary/large-secondary carburetors.

A "true bolt-on" accessory or part is a rare thing in the automotive business, as you already know if you have done much wrenching. Holley engineers did an outstanding job of creating a part to bolt to the manifold and allow retaining all stock equipment, including air cleaner, linkage and *all* emission-control connections.

Initial development was done on Chevrolets, with the 650 CFM unit calibrated on a 350 CID engine and the 800 CFM unit set up as a result of tests on both 402 and 454 CID engines. Under actual driving conditions, the Model 4165 outperformed the original carburetor in traffic accelerations, hot starting, hot-idle conditions, vapor-lock prevention and general driving. In basic wide-open-throttle acceleration tests on a 1971 350 CID automatic transmission Impala, the Model 4165 was 0.85 seconds and 5.4 MPH faster over a 1/4-mile drag strip. Dynamometer testing showed performance improvements of 14 HP on both small and big-block Chevys. Emission levels are similar to the original equipment.

Several tests charted in the accompanying tables show typical improvements gained with the 4165.

Although the first 4165's were designed specifically for Chevrolets, Holley now offers units for Buick, Oldsmobile and Pontiac. Chrysler 340 and 400 CID versions replace the Carter Thermo-Quad on 1971 and later vehicles.

There is also a 650 CFM 4175 model which is similar to the 4165 except it has vacuum-operated secondaries. It is only available with side-hung fuel bowls and a single accelerator pump.

At first glance the carburetor looks like any other Holley four-barrel of the 4150/4160 double-pumper class. Take our word for it—it's not! Nearly every part on the 4165 is special or unique. Very little is the same on the 4165, except for jets, power valves, bowl vent "whistles," needle and seat assemblies and floats.

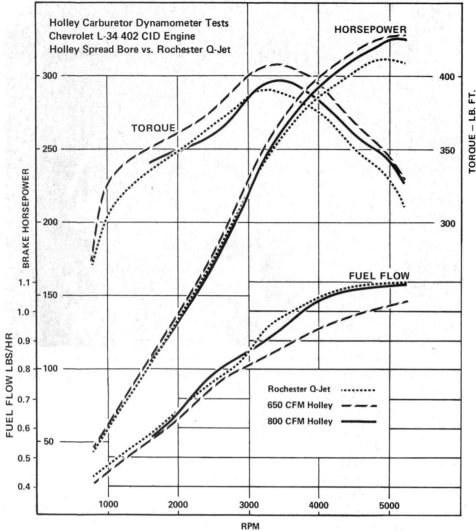

Holley Carburetor Dynamometer Tests
Chevrolet L-34 402 CID Engine
Holley Spread Bore vs. Rochester Q-Jet

HORSEPOWER

TORQUE

FUEL FLOW

Rochester Q-Jet ....
650 CFM Holley - - - -
800 CFM Holley ——————

RPM

Economy and performance seldom come in the same package, but the spread-bore Holley may be an exception from what we can read in these dyno charts. The 650 CFM put out more peak HP and torque—on less fuel, but above 5000 RPM the picture will change. One of these carbs will drop right onto your Q-Jet or Thermo-Quad manifold.

Adjustment and repair of these carburetors is described in the section starting on page 165.

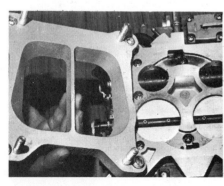

An adapter is used to install any 4165 onto a standard non-spread-bore manifold. This is a Holley adapter which is very efficient. It causes almost no loss of air-flow capacity.

Gaskets, accelerator pumps, metering blocks, pump-discharge nozzles (shooters) and fuel bowls are all different and cannot be interchanged with other Holley parts.

Let's look at the 4165 very carefully to see how it is made and why it is different. First of all, the bolt holes are on a 4-1/4-inch x 5-5/8-inch spacing to fit spread-bore manifolds. Next, depending on the list number, the carburetor has the appropriate connections for emission equipment and the correct linkage to fit the specific automobile using that list number. Some 4165's for street/competition/off-road use where emission controls are not needed may not allow hooking up all emission devices.

One important difference between the 4165 and original-equipment carburetors is the use of fixed-size secondary venturis and main jets. Primary 1-5/32-inch diameter venturis provide good response during part-throttle operation and allow a wide range of vehicle speeds and load conditions without operating the secondaries. Keeping "out of the secondaries," achieves greater fuel economy.

Primary venturis are smaller than the secondaries to allow meeting emissions requirements. A small (20 cc per 10 strokes) accelerator pump on the primaries matches primary-venturi air-flow requirements.

Secondary venturis are either 1-3/8-inch or 1-23/32-inch diameter, depending on whether the carburetor is a 650 or an 800 CFM unit.

Secondaries are mechanically operated and use a high-capacity (50 cc per 10 strokes) accelerator pump.

Accelerator-pump circuits are different on the 4165. This difference makes the 4165 bowl gaskets, metering blocks, fuel bowls and pump inlets and discharge nozzles unlike some other Holley carburetors. Pump circuits are the same for primary and secondary sides of the carburetor. Rubber type inlet/check valves are used. A steel-ball inlet/check was used on the secondary side of some early 4165's.

This difference in pumps between the 4165 and other Holley four barrels resulted from the extensive testing program required to get good hot-fuel-handling characteristics. When the carburetor and fuel are hot, there is always the possibility that there will be an off-idle bog because the pump has "dried out." All pumps have a problem of fuel boiling in the pump cavity (quite close to the engine and subject to a lot of heat) during a hot soak. Several features were incorporated in the design to enhance hot-start driveaway capabilities. Both pumps (primary and secondary) are self-filling so they always fill immediately after the pumps have been used. It is very important for the primary pump to provide an instant shot of fuel at the slightest throttle movement. Because the usual hanging-ball-inlet/check requires an instant to seat before the pump can deliver its shot, Holley chose the rubber-type inlet/check valve.

With this pump arrangement, the inlet/check valve seals instantly as the pump is operated by only a very small movement of the throttle lever. The discharge check is a steel ball near the bottom of the pump passage. This check valve, in conjunction with an anti-pullover discharge nozzle (which prevents fuel in the passage connected to the nozzle from being pulled into the airstream), tends to keep the passage to the "shooter" (nozzle) filled with liquid fuel. As the throttle is moved back toward a closed position, the ball discharge check seats instantly to create a partial vacuum in the pump cavity which further ensures filling the pump quickly. The lightweight ball provides another advantage—it allows vapors to escape from the pump cavity during hot-soak situations where the carburetor heats up when the engine is stopped or idling for a long period. As in many of the other Holley carburetors, discharge nozzles are targeted so the pump shot "breaks" against the lower edge of the booster venturis—ensuring at least partially vaporized fuel entering the engine.

Some 4165's have "reverse" idle adjustment as described and illustrated in the 4150/4160 section. The adjustment screws in to richen the mixture and backs out to lean it—just opposite of what has always been done in the past. Labels identify carburetors with this changed idle system.

There are two types of 4165 carburetors: Standard and high-performance. "Standard," "normal-replacement," or "emission-design/street-performance" 4165 carburetors are really "High-Performance" in so many respects that it seems strange to give them a name which sounds "low-performance." These carburetors

Model 4165 with center-pivot fuel bowls on a 1973 Pontiac 455 CID engine. 800 CFM units similar to this one are available to fit Chevrolet, Chrysler, Pontiac and Oldsmobile engines (except Toronado). **Photo by Mark Campbell.**

DRAG STRIP PERFORMANCE COMPARISONS
Holley 4165 versus Original Equipment

Carburetor	1969 Camaro		1970 Chevelle	
	E. T.	MPH	E. T.	MPH
Original Equipment	14.95	93.01	13.54	104.32
	5 runs average		3 runs average	
650 CFM 4165	14.35	96.05	13.34	105.51
	7 runs average		7 runs average	
800 CFM 4165	14.61	95.44		
	7 runs average			

NOTE: 1969 Camaro equipped with 350 CID engine, Turbo Hydra-Matic transmission, 4.11 gear and open headers.

1970 Chevelle equipped with 350 HP/402 CID engine, Turbo-Hydra-Matic transmission, 3.31 gear and open headers.

Car Craft Magazine Tests — 1970

automatically give a performance increase while holding emissions at legal levels, and offer mechanical secondaries and double accelerator pumps in the same package.

Standard units have a single fuel inlet on the front float bowl. Side-hung float bowls are connected by a fuel-transfer tube which is O-ring-sealed where it enters each fuel bowl. Float settings for these units are accomplished by removing the fuel bowls and bending a tab on each float if resetting is required. A brass baffle on the metering block in the primary bowl reduces the possibility of fuel sloshing through the vent into the carburetor on acceleration. Fuel bowls on the standard carburetors are designed to allow use with stock non-high-performance air cleaners. If the air cleaner has to be rotated slightly for installation, some minor modifying of the air-cleaner-base locating tabs may also be needed.

Automatic chokes are provided on standard 4165's. Buick, Chevrolet and Pontiac units through 1972 use the stock divorced-choke actuating mechanism; 1973 and later Pontiac and Oldsmobile units use an integral choke with an accessory stove with connecting tubing which is installed on the intake manifold heat riser on some models. Chrysler units also work with the stock divorced-choke equipment.

High-performance "Street/competition/off-road non-emission" 4165's are exactly like the "standard" units in every respect except the fuel bowls are the center-pivot-float type with externally-adjustable sight plugs on the side of the bowls for setting fuel level. Two other differences are in the use of "whistle" type bowl vents in both primary and secondary bowls and a manual choke linkage. If you buy a high-performance version you may have to replace the stock air cleaner with a high-performance unit to clear the center-pivot "race" bowls. An accessory fuel line is needed to plumb fuel to both bowls and you must hook up a manual choke to complete the installation. Linkage on these carburetors fits the cars for which each carburetor is designed.

650 CFM versions are recommended for all normal-replacement applications from 327 through 402 CID engines. The same carburetors also work very well on low-performance "smog" version 427 and

454 CID engines. 800 CFM units are recommended for standard replacement on high-performance 396, 402, 427 and 454 engines. They can also be used on racing-type small-block engines with a real need for this much air flow capacity.

In 1974 Holley introduced the Model 4175. It was designed specifically for pick-up trucks, campers, race-car tow vehicles and other applications of the 340 CID and larger engines where low-speed "lugging" makes mechanical secondaries undesirable. Flow capacity is 650 CFM. Divorced and integral chokes are supplied so the carburetor works with the stock equipment on Chevrolet, Oldsmobile, Pontiac and Chrysler applications. It is exactly the same as the Model 4165 except it has diaphragm-operated secondaries. There is no accelerator pump on the secondary side. A metering plate is used on the secondary side.

Top and bottom views of Model 4175, List 6926 intended for Chevrolet trucks. Note vacuum secondary and single accelerator pump. Photos by Tom Stuck.

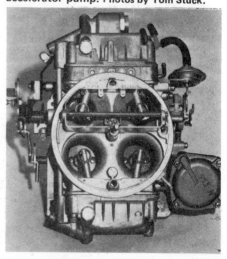

Comparison of 4165 metering block (also 4175 primary) at bottom with 4150 block shows housing for accelerator-pump discharge check ball as indicated by pen. Also, 4165 block has two aluminum plugs on underside.

Pipe plug in hose fitting must be removed before installing line from gulp valve on Chevrolets through 1970 models. Those 4165's without a hose fitting have a manifold-vacuum port in this location for the power-brake line.

Buy your 4165 with the type of fuel bowl you will be needing because it costs money to change to a different type of fuel bowl. 4150 fuel bowls will not work on the 4165 or 4175. Secondary side of two 4165's shown here clearly shows how side-hung bowl is shaped to clear stock air cleaners. The same bowls are used on 4175's for cleaner clearance.

Bowl side of cutaway 4165 metering block shows unique accelerator-pump circuit with ball check at bottom of pump-discharge passage. Ball has its own pressed-in brass seat. Two aluminum plugs and the check-valve housing provide a quick identity check for these unique blocks — which cannot be used on other Holleys.

System routing can be easily understood by comparing passages in cutaway block at top and block at bottom. Top view is from bowl side, bottom view is from the side which fits against the main body.

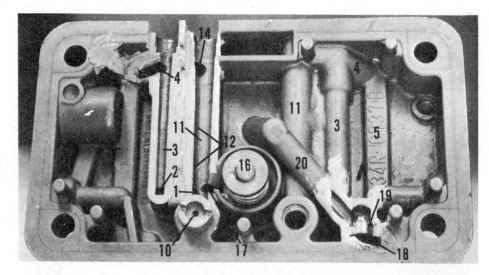

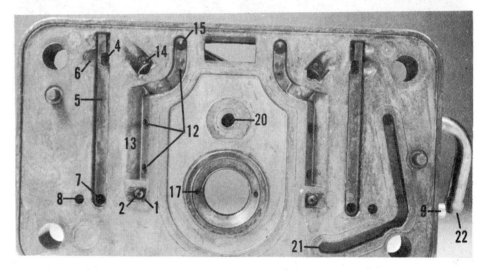

IDLE SYSTEM

1 — Idle feed from main well

2 — Idle-feed restriction to idle well

3 — Idle well

4 — Idle cross-channel

5 — Idle down well

6 — Idle air bleed from main body enters at this point

7 — Idle-transfer fuel is fed to main body from this point. Also feeds curb idle through screw-adjustable needle

8 — Curb-idle discharge passage to main body

9 — Curb-idle mixture-adjustment screw

MAIN SYSTEM

10 — Main jet

11 — Main well

12 — Air-bleed holes from main air well

13 — Main air well

14 — Discharge nozzle passage

15 — Air bleed from main body enters at this point

POWER SYSTEM

16 — Power valve

17 — Power valve channel restriction (connects to main well **11**)

ACCELERATOR-PUMP SYSTEM

18 — Discharge passage connection from pump in fuel bowl

19 — Pump discharge check ball & seat

20 — Pump discharge to nozzles in main body

TIMED SPARK

21 — Connects to main body at this point

22 — Timed spark tube

Model 4165/4175's are different—So do not try swapping parts between 4165/4175 and 4150/4160 or 2300 carburetors, even though their removable fuel bowls make them look similar. We have covered most of the differences in the accompanying text, but let's review to make sure it is all completely clear.

Fuel bowls—4165/4175 accelerator pump/s delivery hole/s mate with the unique metering blocks containing the accelerator-pump discharge valve near the bottom of the block. Side- or center-hung bowls from other models cannot be interchanged. 4165's are available with side-hung bowls which work with the stock air cleaner, and with race-type bowls which require using a performance-type air cleaner. Model 4175's come with side-hung bowls only. Race-type dual-inlet center-pivot-float bowl kits are available for 4165/4175's. Use 85R-5087 for Chrysler or 85R-4893 for GM applications.

Bowl gaskets—4165/4175 bowl gaskets (8BP-1523 on early models without the accelerator-pump tube connection, 8BP-1572 on later models with the tube) match the accelerator-pump-delivery holes in the mating metering blocks. Other gaskets do not fit the gasket-locators on the 4165 metering blocks (and primary metering block of 4175) and will block off the pump-delivery passage.

Hand choke—If a hand choke is required, buy a 4165 already equipped that way because 4165's with automatic integral or divorced chokes do not easily convert to hand-choke operation. 4175's are all equipped for automatic-choke operation. The Carburetor Numerical Listing and Parts Guide in the Holley Performance Parts Catalog shows that most metering-block-to-main-body gaskets, main jets, power valves and inlet needle/seat assemblies are generally interchangeable between 4165/4175 and the 4150/4160 and 2300 carburetors.

Model 4500

Three members of the Model 4500 family: 6464, 4575 and 6214 (left to right). Airflow capacity is 1050, 1050 and 1150 CFM, respectively. A fourth member of the family (not shown) is the 1150 CFM 7320.

Adjustment and repair of these carburetors is similar to that described for the Model 4150 starting on page 165.

End view of 6214 IR type shows 18cc per 10 strokes accelerator pump with the same heavy-duty accelerator-pump linkage usually found only on the 50cc pumps. Center-hung race bowl with externally adjustable needle/seat has dual inlets to allow plumbing fuel lines from either side. These fuel bowls are provided on all of the 4500's.

The first Holley 4500 was designed for Ford in 1969 for use as a single four barrel on NASCAR Stock Car engines. This expanded four barrel mated a sand-cast carburetor body to the basic Holley metering blocks and fuel bowls. Throttle bore and base stud spacings were changed to allow 2-inch-diameter throttle bores giving 1150 CFM air flow capacity. An accelerator pump was used on each fuel bowl, making these one of the first "double pumpers." The carburetors were made exclusively for Ford, using temporary tooling.

In 1970, Holley made the Model 4500 in production as a single four barrel, List No. 4575. In the process of converting the carburetor from handmade units to production items, some air flow was lost. Thus, the 1-11/16-inch venturi versions of the carburetor flow approximately 1050 CFM, depending on die castings, trim dies and other production variations.

List No. 4575 continued its use on NASCAR engines. For a couple of years restrictor rings were pressed into the car-

buretor bores below the throttle plates to reduce air flow thereby "controlling" power and top speeds. Knurled on the outside, the rings had a 1-11/16-inch inside diameter. Later, special restrictor plates between the carburetor and manifold served the same purpose.

The 4500 carburetor should only be used on a manifold designed for this unit with its 5.380-inch-square stud spacing pattern. Any adapter to fit it onto a smaller manifold will reduce the air-flow capability. Tests showed Holley's own adapter reduced flow by 65 CFM. Another adapter lost up to 255 CFM, making a 4500 which test flowed at 1040 CFM flow a mere 785 CFM.

In 1971 Holley released two additional versions of the 4500. One is an isolated/individual runner (I.R.) carburetor with 1-13/16-inch venturis and long boost venturis. The other is a 1-11/16-inch venturi model optimized for use in a 2 x 4 setup on plenum-type race manifolds.

Grandaddy of the 4500 family is this original sand-cast version made specifically for Ford's use in NASCAR racing. Machined venturis were held in place by boost-venturi supports. Note that the accelerator-pump shooters were part of the main body casting.

Saw a 4500 in half and you'll see the rugged linkage which is fully enclosed once the carburetor is bolted to the manifold. This is the 1:1 linkage on a 6214. Small black arrows indicate accelerator-pump discharge passages. Outline arrows point to vent-tube bosses for bowl venting through main body.

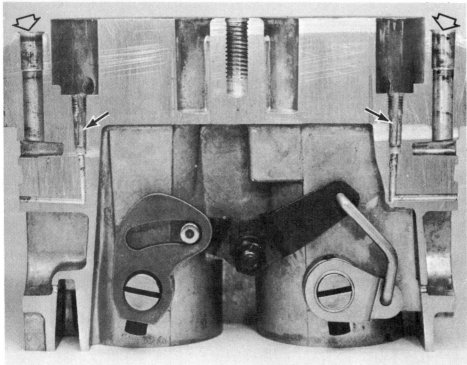

Manifold types are described in the manifold chapter.

1974 saw another 4500—a 1-13/16-inch venturi unit—optimized for use on "mini-plenum" ram manifolds for pro-stock and big-block modified-production engines. This is List 7320.

The four versions are separately described in this section, but a number of features are common throughout the 4500 family. All have 2.00-inch throttle bores and 5.380-inch-square stud pattern. Dual-inlet center-pivot-float fuel bowls are tapped on both sides to allow connecting fuel lines to either side. Plastic bowl vents are provided in all of the fuel bowls. Slabbed reduced-section throttle shafts are used for improved air flow. These 1/2-inch-diameter shafts are supported in ball bearings at each side of the carburetor body. The "blind end" of each shaft is sealed with an expansion plug and the lever end is sealed with an O-ring at the lever end. Linkage between the primary and secondary throttle shafts is enclosed inside the carburetor body to protect it from dirt. Each of the rugged 0.150-inch-thick throttle-shaft levers has a plastic cam to actuate an accelerator pump. Accelerator pumps give 18 cc per 10 strokes units on the 6214 I.R. type; 39 cc per 10 strokes on 4574, 6464 and 7320. Both accelerator pumps are filled through rubber-type inlet valves. These valves seal immediately when the pumps are operated, giving an instant pump shot when there is the slightest movement of the throttle.

Holley does not offer an air cleaner for the 7-1/4-inch mounting pad, but cleaners are made by some of the accessory firms. One open-element cleaner by Sylvania Speed Products showed only a 45 CFM loss at wide-open-throttle, dropping a 1050 CFM carburetor to 1005 CFM total capacity.

Velocity stacks are offered in three- and six-inch lengths. These chrome stacks are typically used on the 6214 and 6464 versions to contain fuel standoff, for tuning purposes—and for their good looks. A slot in the base of each stack allows fuel to pass into the venturi from the pump shooter. Stacks are held in place by a long bolt with a metal safety tab to lock the bolt against loosening. Dynamometer tests showed approximately 3% torque gain with the 3-inch stacks on a 6214

Ford's 427-CID NASCAR engines used the Holley 4500 carburetor.

Ford's "Boss" 429-CID NASCAR engines were also 4500-equipped. Note canister-type filter between mechanical fuel pump and carburetor.

on an isolated-runner manifold, but no flow gain is shown by air-box tests at Holley.

List No. 4575—This single four-barrel 4500 has four 1-11/16-inch venturis fed by number 84 main jets and 6.5-inch power valves on each side (primary and secondary). Air flow is 1050 CFM. Two high-capacity accelerator pumps are actuated by yellow cams. Secondary throttles are operated progressively by a mechanical link to the primary throttle shaft. Connections are provided for timed spark and for a PCV valve. A hand choke is built into the carburetor. If velocity stacks are used, the choke has to be removed.

Operation of the 4575 metering systems is the same as described for the 4150 in that section.

List No. 6214—Isolated/individual runner (I.R.) manifolding supplies each engine cylinder with its own single-barrel carburetor—in this case—one barrel of a Model 4500. The approach is certainly not new because it has been done with small high-performance engines for years—especially on formula racing car engines and motorcycle engines. In fact, some of the first 4500's supplied to Ford were used in an I.R. configuration on Ford 302 Trans Am engines for the 1969 season.

An I.R. system allows ram tuning of the intake system. Engine speed at which the ram effect is most pronounced varies inversely with tuned length. Therefore, the shorter the tuned length, the higher the RPM at which peak torque will occur. The tuned length at wide-open-throttle can be considered to be the distance from the venturi throat (or from the carburetor inlet) to the intake valve. Either of these situations may be true. If venturi size is significantly smaller than the passage *between* the venturi and the valve, the venturi forms a reflection point and defines a tuned length of pipe.

If the venturi is about the same size as the passage between the venturi and the valve, then it does not form a reflection point and the tuned pipe length is determined by the next large change in section outward—probably the air horn or carburetor inlet. For a well-shaped air horn, reflection point will be about halfway along the bellmouth entry. There may be multiple reflections and *two* tuned lengths—one measured from valve to venturi—and the other to the bellmouth entry of the carburetor air horn or velocity stack.

This torque peak can be modified by valve lift, overlap, and by adding a plenum chamber between the runners

and the carburetors. Using a plenum under the carburetors tends to lower the torque peak which can be achieved, but it also broadens the torque peak over a wider RPM range, making the engine less "peaky" and easier to tune. Plenumizing also reduces fuel standoff and allows each cylinder to draw additional mixture from the other barrels at the top end of the RPM range, thereby reducing the carburetor air-flow-capacity requirement for the engine.

Holley's I.R. version of the 4500 is List No.6214 with an air-flow capacity of approximately 1150 CFM through its four 1-13/16-inch venturis. Boost venturis are very long (2-1/4-inch) to contain fuel standoff. Standoff occurs as the fuel/air column starts and stops in the manifold in response to the opening and closing of the intake valve. The pulsing or bouncing movement of the column actually pushes fuel out of the inlet air horns of the carburetor so it can be seen standing above the carburetor as a vapor cloud, hence the term "standoff." The fuel is drawn back into the carburetor on the next intake stroke—but more bounces back out of the carburetor when the intake valve closes. The air column actually pulls fuel out of the discharge nozzle in both directions—on its way into the engine—and as it

Metering blocks for 6214 and 6464 with intermediate idle systems contain no power valves or provisions for installing such. Trying to figure out what systems are fed by which passages has baffled the experts on more than one occasion. This cutaway block should help those who want to know what's happening. Main well 1 contains idle tube 2 with an idle-feed restriction in its tip 3. Fuel is pulled up the tube, through the horizontal cross channel 4, mixing with air from the air bleed at 5. Idle down leg 6 connects at hole 7 to transfer passage and to adjusting needle 8 which varies mixture passing into curb-idle hole 9. Passage 10 feeds the intermediate-idle-system discharge nozzle. 11 is the point where main-air-bleed passage in main body connects to main air well in metering block. 12 is the point where intermediate idle system air bleed enters from main body.

Bowl side of same metering block shows intermediate-idle-system details: Bowl-fed restriction 1 meters fuel into system well 2. Cross channel 3 meets air bleed at 4. Down leg 5 connects to discharge-nozzle passage 6.

NOTE: The metering-block gasket for 6214 and 6464 carburetors, 8R-1522 can be purchased from Holley dealers or custom made from a standard 4150 metering-block gasket. Place the gasket being replaced against the gasket being modified to see where the eight additional holes have to be cut or punched.

bounces back to the carburetor inlet.

Sharp pulses which occur in an I.R. intake system start main metering operation very quickly so a large accelerator pump is not required to cover up nozzle "lag." The accelerator pumps are small units operated by white plastic cams on the throttle levers. There is no provision for a power valve in the 6214 metering blocks because a power valve could not function correctly under the extreme pulsing conditions which exist in an I.R. manifold. Main jets are number 75's for true I.R. manifolds; 95's for plenum-type manifolds (the smaller the plenum volume, the smaller the jets). Throttles open simultaneously

with 1:1 linkage and there are four separate idle systems—one for each barrel. No choke is provided because the carburetor is designed strictly for racing applications.

The main metering and accelerator-pump systems of the 6214 function like those described for the Model 4150 in that section. Idle and intermediate systems function as described in the basic section on how your carburetor works.

The 6214 is not designed for use on *any* single four-barrel manifold. It is not for use on street-driven vehicles.

List No. 6464—Features of the 4575 and 6214 versions are combined in this unit which was developed specifically for pro-stock racing with large engines and deep-

plenum-type manifolding mounting two of these carburetors. Air flow is approximately 1050 CFM. There are four 1-11/16-inch venturis, each with short booster venturis (plenumizing practically eliminates fuel standoff). Two high-capacity accelerator pumps are operated by yellow cams on the throttle levers. Throttles open simultaneously with 1:1 linkage.

Metering blocks are the same as those used in the 6214 with no power-valve provisions and individual idle-mixture screws for each of the throttle barrels. No. 88 main jets are installed in the metering blocks.

No choke is provided because this is strictly a racing carburetor.

The main-metering and accelerator-pump systems of the 6464 function like those described in the section on the Model 4150 carburetor. Idle and intermediate-system function as described in the basic section on how your carburetor works.

The 6464 is not designed for single four-barrel use on *any* manifold. It is not for use on street-driven vehicles.

SPECIAL METERING BLOCKS FOR 6214 and 6464

Racers found the carburetors would not flow sufficient fuel and engines loaded up coming off idle. Holley made a metering block kit 85R-5230 to replace the stock metering on 6214 and 6464 carburetors used on plenum-type manifolds for off-road racing. These metering blocks can flow more fuel through jets and a power valve than the original blocks could flow through the jets alone. Under no circumstances should these blocks be used on any other list number carburetor or on applications that are to be street-driven or requiring part-throttle operation for extended periods.

The following jets will produce a wide-open-throttle mixture identical to the stock jetting of these carburetors when used with a 25BP-591-XX power valve: 75 main jets for the 6214; 65 main jets for the 6464.

Because of the intermediate-idle system used on these two carburetors, they cannot be modified—even with the special metering blocks—to flow as much fuel as a stock 7320.

List No. 7320—Intended for 2 x 4 "mini-plenum-ram" engines for Pro Stock and Modified Production engines, this carburetor can also be used for single four-barrel applications such as the NHRA Econo-Rail and some AHRA Super-Stock classes. The carburetor has 1-13/16-inch venturis and 2-inch throttle bores yielding 1150 CFM air flow. Throttle linkage is progressive.

Dan Parker and family's B/A roadster utilizes a Pro-Stock Hemi with two Model 4500 List 7320's and Holley P-6230 electric fuel pump. A relatively competitive car with fuel injection, Dan lowered his 1/4-mile E.T. by 0.2 second when he switched to carburetors. Photo by Mark Campbell.

The idle-transfer system is unique and has an idle-mixture adjustment screw for each bore. There is no intermediate-idle system. A new booster venturi, more efficient at low air flows, is used. Fuel passages are slightly larger for greater flow. High-capacity accelerator pumps are used with yellow cams. There is no choke.

Power-valve-restriction passages are machined but power-valve plugs are inserted. Two purposes are served here. On 2 x 4 applications, if greater than size 100 main jets are required, install a power valve to supply extra fuel. Secondly, a power valve should be installed for single 4-barrel applications.

DESCRIPTION OF OPERATION

No description of operation is provided for the **4500** because these are essentially "stretched" versions of the 4150/4160 carburetor. Only the intermediate-idle system for the 6214 and 6464 is different and this is fully detailed in the idle system portion of the basic section earlier in the book.

Model 4500 Exploded View Identification for drawing on page 68

1	Choke plate
2	Choke shaft assy.
3	Fuel bowl vent baffle drive screw
4	Choke shaft swivel screw
5	Fuel pump cover screw & LW
6	Choke plate screw & LW
7	Throttle stop screw
8	Fuel bowl screw
9	Pump cam lock screw
10	Fuel valve seat lock screw
11	Pump operating adj. screw
12	Pump discharge nozzle screw
13	Float shaft bracket screw & LW
14	Throttle shaft screw
15	Throttle plate screw
16	Pivot screw
17	Fuel level check plug
18	Fuel inlet plug
19	Fuel level check plug gasket
20	Fuel bowl screw gasket
21	Power valve gasket
22	Fuel valve seat adj. nut gasket
23	Fuel valve seat lock screw gasket
24	Pump discharge nozzle gasket
25	Fuel bowl gasket
26	Metering body gasket — pri. & sec.
27	Fuel inlet fitting & plug gasket
28	Flange gasket
29	Throttle plate
30	Throttle shaft assy. — primary
31	Throttle shaft assy. — secondary
32	Primary throttle lever (internal)

Model 4500 Exploded View
Holley Typical View 42—1

NOTE — General view is useful for visualizing relationship of various parts in the carburetor. Specific details will vary with Part Numbers because each carburetor is made to fit a particular application.

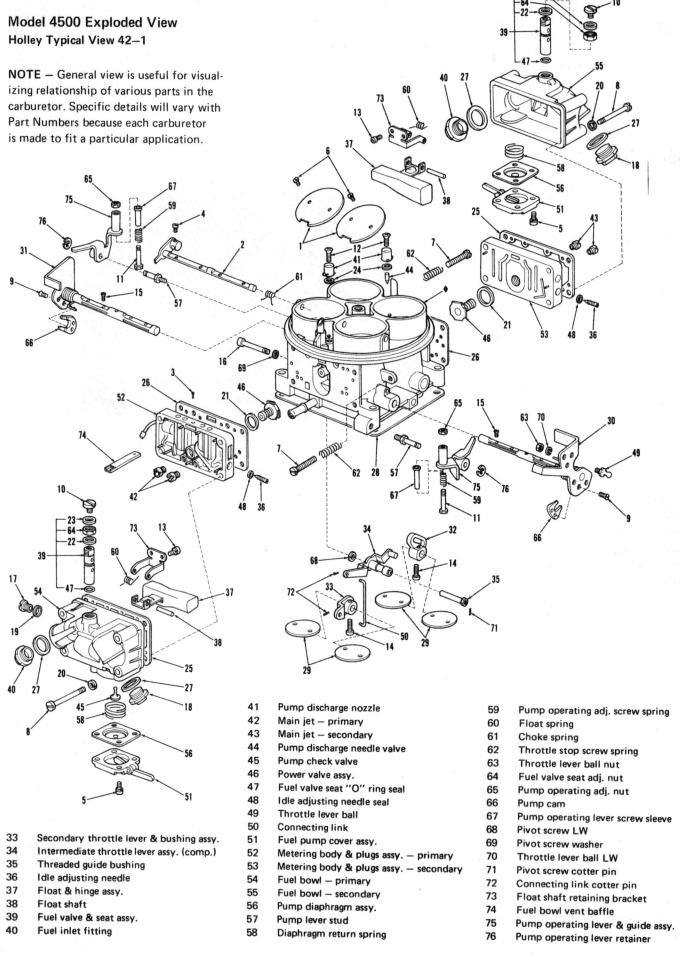

33	Secondary throttle lever & bushing assy.
34	Intermediate throttle lever assy. (comp.)
35	Threaded guide bushing
36	Idle adjusting needle
37	Float & hinge assy.
38	Float shaft
39	Fuel valve & seat assy.
40	Fuel inlet fitting

41	Pump discharge nozzle		59	Pump operating adj. screw spring
42	Main jet — primary		60	Float spring
43	Main jet — secondary		61	Choke spring
44	Pump discharge needle valve		62	Throttle stop screw spring
45	Pump check valve		63	Throttle lever ball nut
46	Power valve assy.		64	Fuel valve seat adj. nut
47	Fuel valve seat "O" ring seal		65	Pump operating adj. nut
48	Idle adjusting needle seal		66	Pump cam
49	Throttle lever ball		67	Pump operating lever screw sleeve
50	Connecting link		68	Pivot screw LW
51	Fuel pump cover assy.		69	Pivot screw washer
52	Metering body & plugs assy. — primary		70	Throttle lever ball LW
53	Metering body & plugs assy. — secondary		71	Pivot screw cotter pin
54	Fuel bowl — primary		72	Connecting link cotter pin
55	Fuel bowl — secondary		73	Float shaft retaining bracket
56	Pump diaphragm assy.		74	Fuel bowl vent baffle
57	Pump lever stud		75	Pump operating lever & guide assy.
58	Diaphragm return spring		76	Pump operating lever retainer

Model 4360

Model 4360: Compact design, light weight, low price. Fuel inlet (F) allows the use of the original fuel-supply line. Single tubes are: manifold vacuum 1, EGR signal 2, canister purge 3, PCV 4, and distributor or "spark" vacuum 5.

Adjustment and repair of these carburetors is described in the section starting on page 203.

This compact, lightweight spread-bore carburetor was introduced in 1976 as a cost-competitive replacement for the Rochester Quadrajet. Initial list numbers for the 1965-74 Chevrolets and 1968-72 Pontiacs will be supplemented with other GM and Chrysler applications.

One objective of the design was to provide a carburetor fitting the trend toward smaller engines, as seen in GM's 260, 262, and 305 CID V-8's and GM and Ford V-6's. Here the Model 4360 serves as an excellent *performance* carburetor because the small primaries give excellent throttle response and feel in the low and mid-ranges.

The combination of small venturis and a good delivery system yields excellent atomization and strong signals even at low air flows. These manifest themselves in good drivability and fuel economy and low exhaust emissions.

The mechanical secondaries are available for acceleration or top power output when required. The carburetor also works as a fine normal-driving replacement for big-block engines.

BRIEF SPECIFICATIONS

Primary bore size is 1-3/8; the secondary is 1-7/16. This combination yields a total airflow of 450 CFM tested at wide-open throttle with a 1.5 inch Hg. pressure drop. About 40% of the flow comes through the primary side. At first glance 450 CFM seems a little small, especially if you are intrigued by high-flow carburetors, but when flowed at 3 inch Hg., as two-barrel carburetors are, it flows 630 CFM. This represents about a 60% gain over the largest original-equipment two barrel and is 30 CFM larger than Holley's 6425 giant two-barrel.

A single fuel bowl integral with the

Universal throttle lever satisfies all Chevrolet and Pontiac hook-ups. Primary lever moves alone for 40 degrees until tang (arrow) meets tang on the intermediate free-floating lever to actuate the secondary-throttle lever.

Top of main body with air horn removed. Float is held by combination hinge pin and retainer. Bowl capacity is about 90 cc.

main body nestles between the venturis. To accomplish this, secondary-bore centers were widened 0.36-inch from the Quadrajet. This was possible because the much smaller secondary bores, even with the offset, fit well within the manifold's 2-1/4 inch bores. This unique arrangement is the key to the whole compact design philosophy. Primary bore centers and all stud spacings are the same as the Quadrajet.

The 90cc. fuel bowl has proved more than adequate under severe temperature conditions. The brass float is mounted at the front of the fuel bowl. Fuel enters from the front through a single 0.110-inch inlet valve with a Viton needle. Level adjustment is accomplished by bending the float-hinge tab.

You'll probably recognize the primary main metering system is similar to the one used in the Models 5200 and 5210.

The secondary main system is a diagonal well type formed inside and adjacent to a cast-in booster venturi. This very efficient system helps to get the main system flowing at very low air flows. This, combined with the small venturis, allows the use of mechanical secondaries without adding a pump to the secondary side. They begin to open when primary throttles are at about 40°.

The idle systems are conventional. While both primaries and secondaries have idle systems, only the primary side is adjustable. This adjustment is right at the discharge where there is maximum velocity to help vaporization and cylinder-to-cylinder fuel/air distribution. Adjustment is conventional—lean clockwise and rich counter-clockwise.

All metering restrictions—other than the main jets and power valves—are pressed in. The main jets are part of a new metric-threaded line introduced with this carburetor. They are similar to those in the Models 5200 and 5210 and available in a much broader size range.

The three-piece construction has separate aluminum castings for the air horn, main body and throttle body. Aluminum is lighter than zinc and more stable at higher temperatures, making for tighter sealing over the life of the carburetor. Aluminum is also easy to machine.

The three-piece construction allows more design freedom and the use of heat-insulating gaskets between the castings. The complete assembly weighs only 6.4 lbs. as compared to 8.8 lbs. for the Quadrajet. It is only 1/2 pound heavier than the Rochester two-barrel.

The spring-driven piston-type accelerator pump has a synthetic-rubber cup. Three adjustment positions are provided.

When this design was started, many emissions features, such as the EGR signal system, evaporative vent valve, throttle-control solenoids and others were already defined. Rather than add these features as had to be done on existing models, they were incorporated in the original package. All necessary vacuum tubes and ports are provided so this carburetor is a true "bolt-on" replacement

The choke on the Chevrolet and Pontiac types is of the remote or divorced variety with a diaphragm-type qualifying system. The original choke rod and sensing unit can be used. Integral choke designs will be provided to fit other applications.

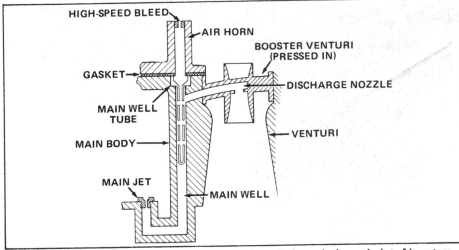

PRIMARY MAIN SYSTEM. Fuel enters the main well through the main jet. Air enters main well through the high-speed bleed and the holes in the well tube. The air and fuel are mixed forming an emulsion which is discharged into the air section through the booster venturi.

SECONDARY MAIN SYSTEM. Fuel enters the main well from the main jet and flows up the diagonal well where it meets the air entering through the high-speed bleed. The emulsion is then discharged through the booster-venturi nozzle.

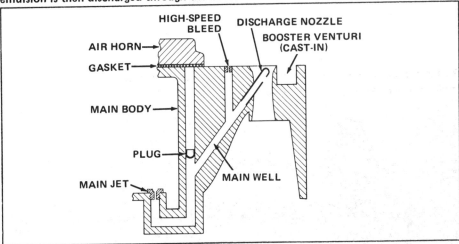

Cast-in secondary booster showing high speed bleed 1 and discharge nozzle 2. Idle feed 3 and tracking similar to the primary side.

Three principal castings and their assemblies, air horn, main body and throttle body.

View of the air horn showing primary idle air bleeds 1, primary high speed bleeds 2 and secondary idle air bleeds 3.

Idle-mixture screw 1 and idle-speed screw 2. Some list numbers have a plastic limiter cap on the idle mixture screw.

Pressed-in primary nozzle or booster venturi 1. 2 shows primary idle-feed restriction. Idle fuel meets the idle-air bleed in a matching pocket in the air horn. The emulsion travels through the track 3 and down a diagonal passage to the discharge. Idle-channel restriction 4 is left.

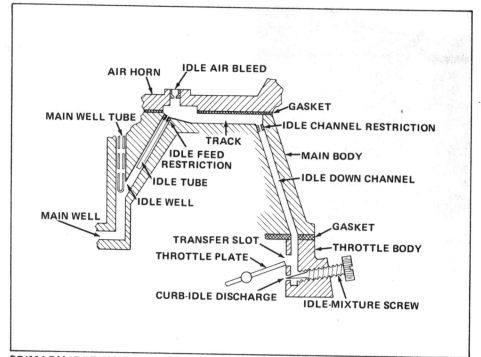

PRIMARY IDLE SYSTEM. Idle fuel is drawn out of the main well and is metered by feed restriction at the top of the idle tube. Air enters at the idle air bleed just above the feed restriction and the emulsion moves along a track in the top of the main body, then through a channel restriction and finally down the channel to a mating passage in the throttle body. The throttle body contains both an adjustable discharge and a transfer slot. The idle mixture control is conventional, clockwise to lean and counter-clockwise to enrichen.

SECONDARY IDLE SYSTEM. Idle fuel is drawn up the idle tube from the main well and metered by the restriction at the top. The metered fuel travels along the track in the top of the main body and meets air from the idle air bleed at the top of the vertical channel. The emulsion travels down the channel and is discharged out the transfer slot in the throttle body. The design also provides for a non-adjustable discharge port below the throttle plate.

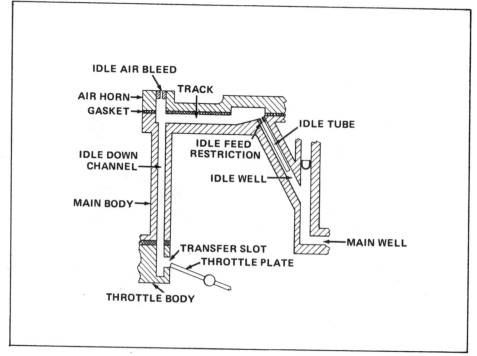

Choke lever hooks up to original choke rod. Hole 1 is for units that pull to close. Hole 2 is for those that push to close.

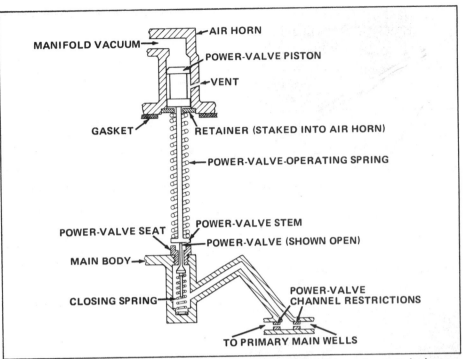

POWER ENRICHMENT SYSTEM. This simple single-stage enrichment system is shown in the rich or "on" position. Manifold vacuum tries to pull the power-valve piston up and the operating spring tries to pull it down. When manifold vacuum drops to a given value, the spring forces the piston and stem down and opens the power valve in the bottom of the fuel bowl. Fuel flows through the valve and on to power-valve-channel restrictions where it is metered prior to entering each primary main well.

ACCELERATOR PUMP SYSTEM. Fuel is drawn into the pump well through the floating cup as it moves up responding to the closing of the throttle. While this is happening the outlet check seals, preventing air from entering the well. As the throttle is opened, the drive spring forces the piston down. This, in turn, forces fuel past the outlet check and through the shooters into the primary air section. The kill bleed arrangement prevents fuel pull-over at high air velocities.

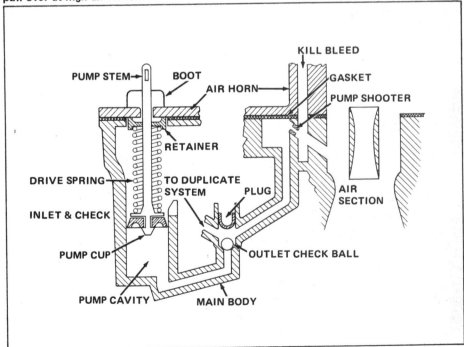

MODEL 4360 PERFORMANCE

Performance tests were run at Holley on an engine dynamometer. The emission test laboratory was also used for checking emissions. Emission test procedures have changed over the years, so the tests performed related to the model year of the test vehicle.

Economy data is from the EPA city and highway test procedures. In some cases, hot emission tests were used for immediate back-to-back comparisons. Because this omitted the required 10-hour soak period, this data cannot be compared to federal standards.

The accompanying charts provide power, torque and emission performance comparisons.

The H.P. chart shows Holley dynamometer data from a new 1975 Chevrolet LM-1 350 CID engine illustrates torque and horsepower comparison between Holley 4360 and the OEM two-barrel and the OEM four-barrel Quadrajet. All data was obtained within a time span of two days on the same engine with an intake-manifold change to run the two-barrel. Peak torque of the Model 4360 occurred at 2800 RPM and exceeded both OEM carburetors. At the high end, both the torque and horsepower curves of the 4360 fall between the 2V and Q-J as you would expect when comparing air-flow capacity of the carburetors. Note how very close it is to the much larger Q-J. Uniformity of the fuel/air mixture between cylinders was within ± 9% across all the cylinders at all W.O.T. points (within one F/A ratio). Model 4360 was designed to fulfill the need for a small carburetor which drives well and performs better than the much larger Q-Jet. In addition, it offers a performance option to those who wish to convert their two-barrel installation to a four-barrel. Many owners want to do this to make today's smaller and more congested engines run better.

Illustrated here is comparative data on fuel economy and tailpipe emissions between the 4360 and the Q-Jet, using the 1973 EPA hot-test procedure. Hydrocarbon and carbon-monoxide emissions for the 4360 were considerably lower than those obtained with the OEM carburetor. At the same time fuel economy was noticeably improved.

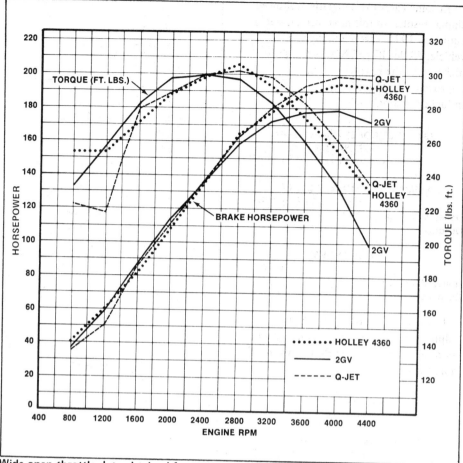

Wide-open throttle data obtained from a new 1975 Chevrolet LM-1 350 CID engine compares the Holley 4360 power and torque with that available from the Quadrajet and 2GV Rochester carburetors.

1972 CHEVROLET MONTE CARLO
350 CID Engine
4-V Comparisons

EMISSIONS '73 E.P.A. (HOT) GM/MI

	H.C.	C.O.	NOx
Holley Model 4360	.91	8.05	2.26
Quadrajet	1.76	18.93	2.25

FUEL ECONOMY
HOT E.P.A.

	(City)	HIGHWAY
Holley Model 4360	16.7	24.7
Quadrajet	16.4	23.5

Comparison between the 4360 and stock two-barrel carburetor run on the same car after a manifold change shows all tailpipe emissions were significantly reduced. Gasoline mileage was increased. Makes you wonder about the myth that four-barrel carburetors give poorer fuel economy than two-barrels.

1972 CHEVROLET MONTE CARLO
350 CID Engine
4-barrel vs. 2-barrel

EMISSIONS '73 E.P.A. (HOT)

	H.C.	C.O.	NOx
Holley Model 4360	.91	8.05	2.26
2GV Rochester	1.64	21.15	2.85

FUEL ECONOMY
HOT E.P.A.

	(City)	HIGHWAY
Holley Model 4360	16.7	24.7
2GV Rochester	15.1	22.4

In some cases the comparison showed similar results. In this next test a Q-Jet equipped 1971 wagon was the test vehicle. While HC and NO$_x$ are slightly higher with the Holley, CO is greatly improved. The Q-Jet has a slight edge in fuel economy.

1971 CHEVROLET KINGSWOOD ESTATE WAGON
402 CID Engine

EMISSIONS 7 MODE 7 CYCLE

	H.C.	C.O.	NO$_x$
Holley Model 4360	3.04	10.52	2.96
Quadrajet	2.97	36.44	2.72

FUEL ECONOMY

	HIGHWAY
Holley Model 4360	21.7
Quadrajet	21.8

In the next test on a 1974 Nova, tailpipe emissions and fuel economy are essentially equal but the Holley was much more throttle responsive.

1974 CHEVROLET NOVA
350 CID Engine

EMISSIONS '73 E.P.A. (HOT)

	H.C.	C.O.	NO$_x$
Holley Model 4360	1.57	22.12	1.88
Quadrajet	1.87	16.05	1.63

FUEL ECONOMY

	HOT E.P.A. (City)	HIGHWAY
Holley Model 4360	17.5	22.5
Quadrajet	17.9	22.5

Drivability is a very subjective thing and it's hard to put a number on it, but it is our opinion that the little Holley four-barrel offers a plus in almost every area.

NEW MAIN JET SERIES

With the introduction of the Model 4360 in 1976, Holley started to use a new series of main jets with tighter flow tolerances. These jets are dyed green to identify them. The number stamped on the jet relates to the average flow range of the jet. Example: A 22R-130A-203 (stamped 203) flows approximately 203 cubic centimeters per minute with a head of 50 centimeters.

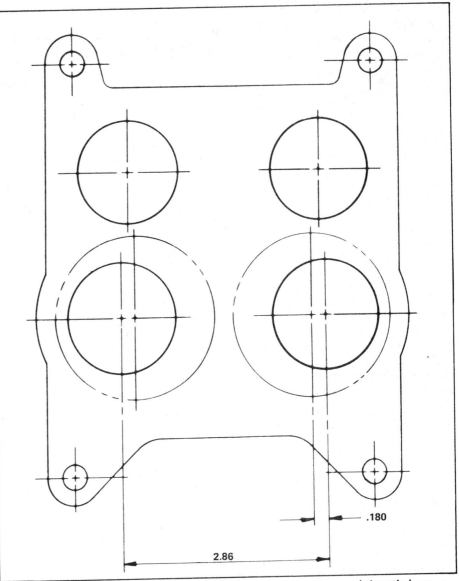

Flange configuration illustrates the Quadrajet flange bolt locations and throttle bore spacing superimposed over that of the 4360. As you can see, the flange bolts and primary bores of both carburetors are located on exactly the same centerlines. Primary bores are the same size at 1-3/8 inch. Model 4360 secondaries are considerably smaller: 1-7/16 inch as compared to 2-1/4 inches. Its secondary bore centers are spread slightly further apart but are well within the 2-1/4 inch circle.

Model 1920 has very few parts and is extremely easy to work on as a logical result. It has been used as the original-equipment (OE) carburetor on slant-six Chrysler engines. Early models (different numbers) were on Ford products.

Adjustment and repair of these carburetors is described in the section starting on page 181.

Two kinds of accelerator-pump actuation are used on Model 1920 carburetors. At left is the cam-drive arrangement. Spring-driven type is at right.

Model 1920

The little 1920 is possibly one of the simplest of modern-day carburetors. It has very few parts, is extremely reliable and easy to service. The carburetor has been used in great numbers on the popular 170 and 225-CID slant-six Chrysler engines used in Dodge and Plymouth applications. It is also basically the same carburetor as earlier **1904** and **1908** models, many of which were used as original equipment on Ford cars and trucks.

BRIEF SPECIFICATIONS

This single-barrel carburetor is available in flow capacities of 160 and 235 CFM (flow at 3.0 inches Hg). Throttle bores are 1-9/16 and 1-11/16 inch. A 1-1/2-inch SAE flange with 2.94-inch stud centers is used for both types. The airhorn measures 2.3 inches outside diameter for a bail-type air-cleaner mounting.

Three castings are used. The body is aluminum and the fuel bowl and metering block are cast of zinc. The fuel bowl (or bowl cover) is only half a bowl with the rest of the bowl in the carburetor body. Some **1904** models actually used glass for the outer portion of the fuel bowl. The unique metering block contains the main metering system (main jet, main fuel well and main air well with bleed holes for emulsioning), accelerating-pump inlet and discharge check valves, power valve and power-valve-channel restriction and the idle-fuel-feed restriction and well.

A closed-cellular float is side-hung in the fuel bowl. The power valve is either one- or two-stage, depending on the specific vehicle-emission requirements. An idle limiter is used on the idle-adjustment screw. An evaporation-control valve in the top of the fuel bowl vents to an external canister or storage device when the throttle is closed. At off-idle, the stored vapors are drawn into the carburetor air horn.

The accelerator pump may be either spring-driven or cam-driven.

Chokes are offered in manual, divorced and integral types. The divorced choke has a vacuum qualifying diaphragm. Hose connections include PCV, timed spark, manifold vacuum and evaporation control.

DESCRIPTION OF OPERATION

Operation of the 1920 is as described in the earlier section on how your carburetor works.

Model 1920 Systems

Inlet System

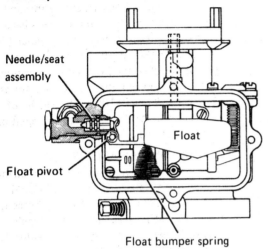

- Needle/seat assembly
- Float pivot
- Float
- Float bumper spring

Accelerator Pump System

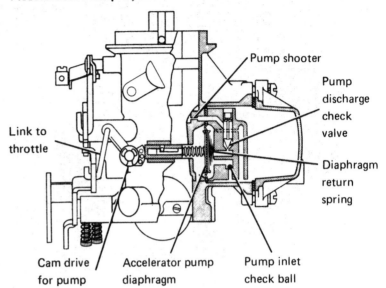

- Link to throttle
- Pump shooter
- Pump discharge check valve
- Diaphragm return spring
- Cam drive for pump
- Accelerator pump diaphragm
- Pump inlet check ball

Idle System

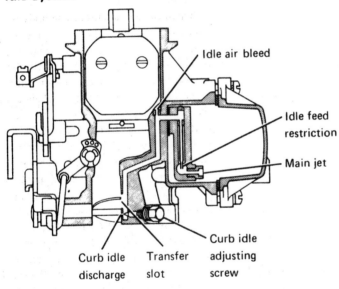

- Idle air bleed
- Idle feed restriction
- Main jet
- Curb idle discharge
- Transfer slot
- Curb idle adjusting screw

Main System

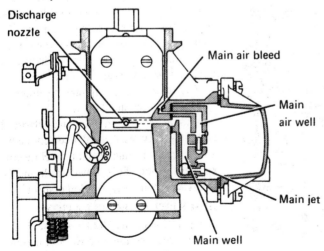

- Discharge nozzle
- Main air bleed
- Main air well
- Main jet
- Main well

Power System

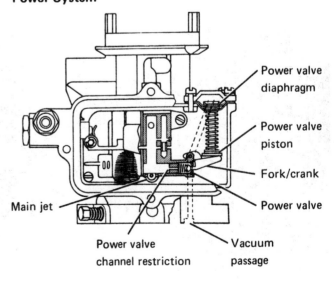

- Power valve diaphragm
- Power valve piston
- Fork/crank
- Power valve
- Main jet
- Power valve channel restriction
- Vacuum passage

Models 1940 & 1945

The Model 1940 has seen wide use on six-cylinder Ford car, truck and industrial engines, usually with an Autolite trademark on the carburetor body. It is also available as a Holley aftermarket replacement for Ford, Chevrolet and other makes.

Model 1945 is very similiar to the 1940 except for the throttle-lever, choke and power-valve. It was first used on the Chrysler Corporation six-cylinder engine in 1974. Because the 1940 is used to describe the systems, we have used the 1945 for the repair and adjustment portion. The 1945 has a divorced choke in this description. The 1940 we've pictured has an integral choke, but some 1940's have divorced chokes.

BRIEF SPECIFICATION

The single-barrel Model 1940 is available in three flow capacities: 170, 180 and 212 CFM (measured at 3.0-inches Hg pressure drop). Throttle bores are 1-7/16, 1-9/16 and 1-11/16-inch. All share a 1-1/2-inch SAE flange. Ford models utilize a 2.68-inch stud spacing in this flange size; Chevrolet versions use a 2.92-inch stud spacing. The air-cleaner flange is 2-3/16-inch diameter. The Model 1945 comes in one size with a 1-11/16-inch throttle bore and flows 203 CFM at a 3.0-inch Hg pressure drop.

Three major castings are used. Air horn/bowl cover and bowl/body castings are zinc and the throttle body is aluminim.

The nearly concentric fuel bowl allows the carburetor to meet military angularity specifications. Hose connections are provided for PCV valve and tapped opening/s for conventional timed spark or spark valve (early Ford vehicles) are included as required. A fresh-air pickup in the air horn supplies filtered air to an integral choke (when used). Teflon coating on the throttle and choke shafts and on a choke lever minimizes friction and provides long-wearing capabilities.

DESCRIPTION OF OPERATION

This description is applicable to all Model 1940 and 1945's. No attempt has been made to explain why certain things happen in the functioning of the carburetor because, except as noted, operation is as described in the earlier section on how your carburetor works.

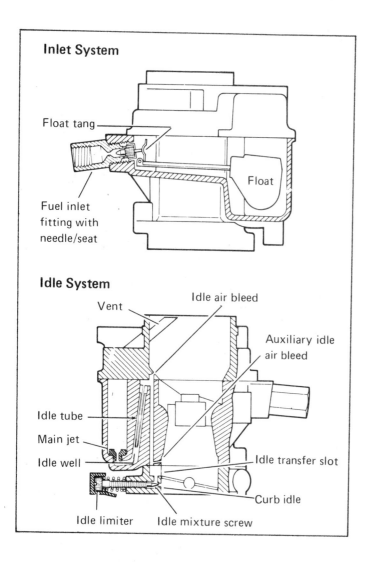

Inlet System

Float tang

Fuel inlet fitting with needle/seat

Float

Idle System

Vent

Idle air bleed

Auxiliary idle air bleed

Idle tube

Main jet

Idle well

Idle transfer slot

Curb idle

Idle limiter

Idle mixture screw

Fuel Inlet System—Fuel enters the fuel bowl through a fuel fitting in the carburetor body. The Viton-tipped needle seats directly in the fuel-inlet fitting, where it is retained by a cap. Fuel flows through holes in the side of the inlet fitting. The fuel bowl is designed so there is no need for a baffle. The float is a dual-lung closed-cellular unit mounted on a stainless-steel float lever hinged with a stainless-steel float pin.

Main Metering System—At cruising speeds, fuel flows from the fuel bowl, through the main jet into the bottom of the main well. The main air bleed in the carburetor cover supplies filtered air to the emulsion tube inserted in the main well. As fuel moves up the main well, it is mixed with air as it passes the holes in the emulsion tube. This mixture of fuel and air moves up the main well and across the discharge nozzle in the dual-booster venturi.

The main discharge nozzle passage is incorporated in the booster venturi support which is an integral part of the carburetor body. Distribution tabs in the main venturi help to ensure equal distribution of the fuel/air mixture to all cylinders when the carburetor is used in conjunction with the original-equipment manifold.

Accelerator-Pump System—The accelerator pump is a spring-driven piston type in a cylinder formed in the bottom of the fuel bowl. The pump cylinder (cavity) fills through the center of the pump cup, loosely held onto the piston stem.

As the throttle lever is moved, the pump link operates through a series of levers and a pump drive spring pushes the pump piston down, seating the pump cup against the pump piston face so fuel can not escape into the bowl.

When the pump is not in operation, vapors or bubbles which form in the pump cavity can escape around the piston stem through the floating piston cup inlet.

Choke System—The automatic choke system (on 1940's with an integral choke) is controlled by a bimetal spring in a choke housing. This is part of the bowl cover.

When the engine starts, manifold vacuum applied to the choke diaphragm through a passage in the carburetor body opens the choke valve to a preset vacuum qualifying position, called *choke kick*.

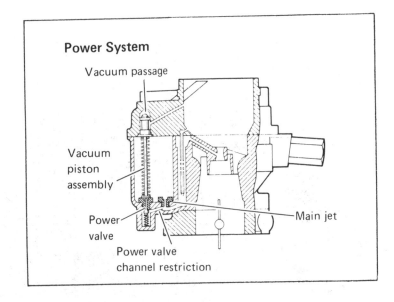

Power System

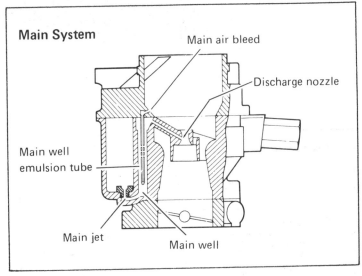

Main System

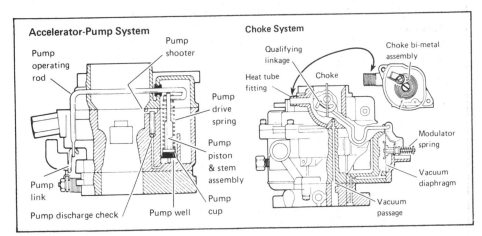

Accelerator-Pump System

Choke System

A modulator spring gives correct initial choke opening after initial start in relation to outside temperature.

Manual choke 1940's use a different air horn casting. Some 1940's and all 1945's have divorced chokes.

Adjustment and repair of these carburetors is described in the section starting on page 186.

Side view of 2210 shows fast-idle linkage and cam, choke vacuum-qualifying diaphragm hose-connected to throttle body, PCV and manifold-vacuum connections. Idle adjustment screw on this Chrysler unit grounds against an insulated stop used for changing the distributor spark advance at idle. Tab on accelerator-pump lever (top left) opens an evaporative control vent at idle so vapors can be collected in an external storage device.

Adjustment and repair of these carburetors is described in the section starting on page 191.

Solenoid is used on some 2210's to provide a curb-idle setting. It allows the throttle to close farther when the ignition is turned off, greatly reducing any dieseling (after-run) tendency.

Models 2210 & 2245

Model 2210 carburetors have been used as original equipment on Chrysler 383 and 360 engines. Aftermarket types are offered for both Chrysler and Chevrolet. The model 2245 was introduced on 1974 Chrysler Products. It differs from the model 2210 principally in the power-valve department.

BRIEF SPECIFICATIONS

The 2210 & 2245 two-barrel carburetors flow 380 CFM at 3.0-inch Hg pressure drop. Throttle bores are 1-9/16-inch diameter in an SAE 1-1/2-inch two-barrel flange with a 2.00 X 3.68-inch stud pattern. The air-cleaner-mounting flange is 4-3/16-inch diameter (adapter supplied for some applications).

Three major castings are used: Zinc air horn and main body and an aluminum throttle body. A triple-venturi design is cast into the main body (two booster venturis plus main venturi) for good atomization of the fuel as it leaves the discharge nozzle. A closed-cellular float is side-hung in the fuel bowl. The accelerator-pump is a spring-driven piston type.

The model 2210 power valve is either one- or two-stage, as required for specific vehicle-emission requirements. The model 2245 has a gradient or smoothly-varying power valve. Idle limiters are used on the idle-adjustment screws. An evaporation-control valve in the cover vents the bowl to an external canister or storage device (crankcase on some of the Chrysler products) and provides purging of the stored vapors into the carburetor when the throttle is moved off-idle.

Idle screws are slanted for accessibility. Teflon-coated throttle and choke shafts reduce friction and provide long life. The divorced choke has a vacuum-qualifying diaphragm unit hose-connected to the throttle body. Hose connections include PCV, timed spark, manifold vacuum and evaporation control.

DESCRIPTION OF OPERATION

Operation of the 2210 is as described in the earlier section on how your carburetor works.

Systems Drawings for Model 2210

Inlet System

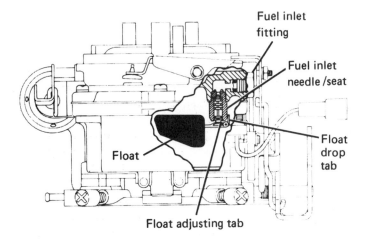

Fuel inlet fitting

Fuel inlet needle /seat

Float drop tab

Float

Float adjusting tab

Power System

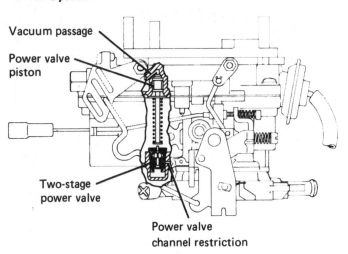

Vacuum passage

Power valve piston

Two-stage power valve

Power valve channel restriction

Idle System

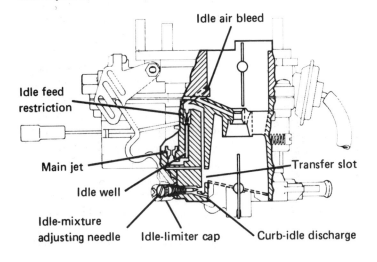

Idle air bleed

Idle feed restriction

Main jet

Idle well

Idle-mixture adjusting needle

Idle-limiter cap

Transfer slot

Curb-idle discharge

Accelerator Pump System

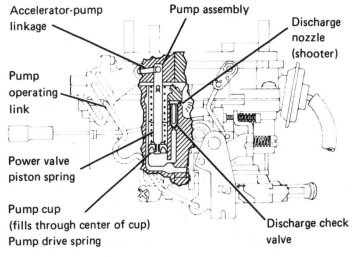

Accelerator-pump linkage

Pump assembly

Discharge nozzle (shooter)

Pump operating link

Power valve piston spring

Pump cup (fills through center of cup) Pump drive spring

Discharge check valve

Main System

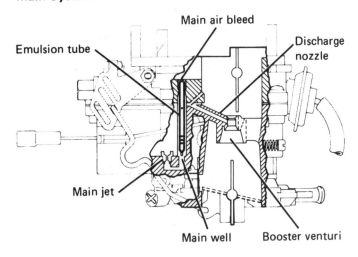

Main air bleed

Emulsion tube

Discharge nozzle

Main jet

Main well

Booster venturi

Choke System

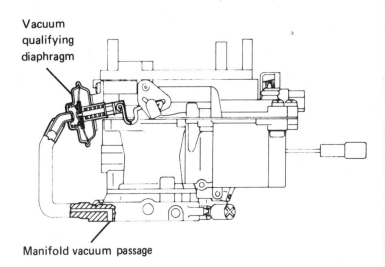

Vacuum qualifying diaphragm

Manifold vacuum passage

Models 5200 & 5210

These two carburetors are high-performance models for small engines. The 5200 is used on Ford products including Pinto, Capri, Mustang II and Cortina. Model 5210 is used on the Vega L-11 engine and on the Pontiac 151 CID four-cylinder engine. These carburetors differ mainly in the throttle-lever and fuel-inlet areas. Also, the 5210 does not have the 5200's deceleration system. Other inconsequential differences exist, but the same description can be used for both. To give you a look at both carburetors, we've described the 5200 systems and used the 5210 for the repair portion.

BRIEF SPECIFICATIONS

These 270 CFM (flow at 3.0 inches Hg) staged two-barrels have a mechanically actuated secondary throttle. Venturi sizes are different as is common for progressive-secondary carburetors used on emission-controlled engines, regardless of size. The primary venturi is 26mm (1-1/32-inch) and the secondary is 27mm (1-1/16-inch). Throttle bores are 32mm (1.25) and 36mm (1.40-inch) diameter, spaced 1.7 inches apart on a metric 2V flange. The rectangular stud pattern measures 1.84 x 3.66 inches.

A diaphragm-type accelerator pump is used. Hose connections are provided for jacket water to the thermostatically operated choke, fuel inlet and return, timed and vacuum spark and for a PCV valve. A connection for the deceleration valve used on Ford products is also provided.

If you have been servicing U.S.-designed carburetors, the 5200's screw-in main air-bleed, main-metering and idle-metering restrictions (jets) may seem unusual.

DESCRIPTION OF OPERATION

This description is applicable to all Model 5200's. No attempt has been made to explain why certain things happen in the functioning of the carburetor because, except as noted, these operate as described in the earlier section on how your carburetor works.

Top photo shows fuel inlet with plug in fuel-return opening, deceleration valve tube (used with external deceleration valve on some Pintos and other Fords), diaphragm accelerator pump and water connections for automatic choke. Lower photo shows choke vacuum-qualifying diaphragm, fast-idle linkage and progressive throttle linkage.

Adjustment and repair of these carburetors is described in the section starting on page 196.

Fuel Inlet System—Fuel under pressure enters the fuel bowl through the fuel-inlet fitting in the air horn (carburetor cover) and through a filter screen. The **5210** has the usual GM-type sintered-bronze filter or a paper filter in the inlet.

The float is a closed-cell structure with two lungs. A small retaining clip hooked over the end of the float-lever tang is attached to the fuel-inlet needle to ensure the fuel-inlet needle will be pulled downward as the float drops. When the float is hanging in its down position, a tang on the float lever contacts a horizontal bumper spring to stabilize the float against undue bouncing.

The fuel bowl is vented to the air horn. On some models (usually for air-conditioned cars) there is a fuel-return connection just above the fuel-inlet. A portion of the fuel supplied by the pump is returned to the tank to ensure liquid fuel is supplied to the carburetor. Any vapors generated in the line between the fuel pump and the carburetor inlet during a hot soak are vented to the fuel tank.

Idle System—There is an idle system for each barrel of the carburetor. The primary side is adjustable and the secondary side is used only as a transfer system to provide mixture as the secondary starts to open and prior to the time the secondary main system starts to operate.

Other than the adjustment on the primary side, the idle systems are essentially identical in construction.

Power System—The primary power system valve is actuated by a diaphragm/rod combination operated by manifold vacuum.

The secondary power system is operated by air velocity through the secondary venturi, which creates a low pressure at the discharge opening in the air horn. Fuel flows from the bowl through a vertical passage containing a restriction. At the top of the passage, air from an air bleed is mixed with the fuel and the fuel/air mixture flows through a cross passage into the air horn.

Accelerator-Pump System—The accelerator pump is a diaphragm type located in the side of the body. It discharges into the primary side of the carburetor only. The pump shooter can be modified to give a pump shot into both sides if desired.

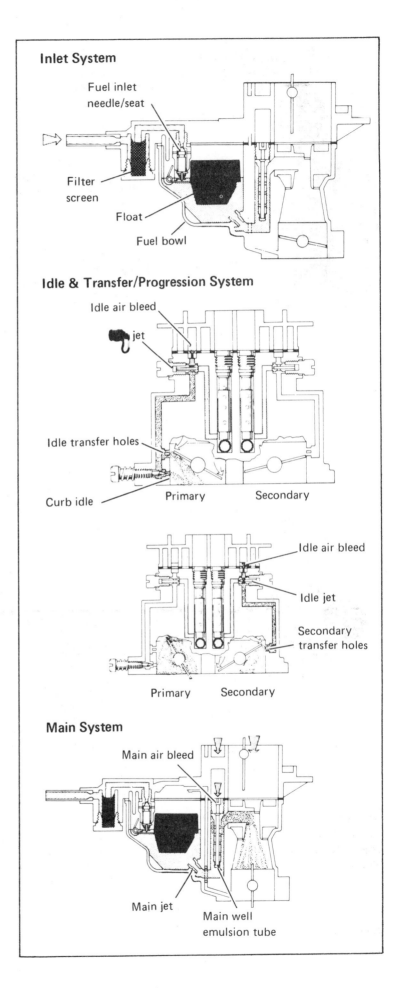

Inlet System

Idle & Transfer/Progression System

Main System

Power System (Primary)

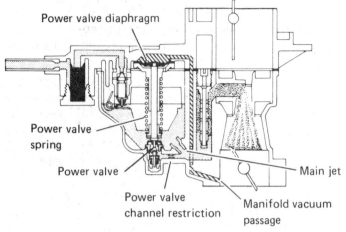

Power valve diaphragm

Power valve spring

Power valve

Power valve channel restriction

Main jet

Manifold vacuum passage

Power System (Secondary)

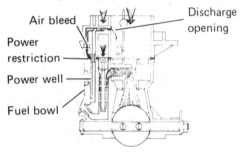

Air bleed

Power restriction

Power well

Fuel bowl

Discharge opening

Deceleration System

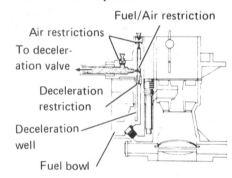

Fuel/Air restriction

Air restrictions

To deceleration valve

Deceleration restriction

Deceleration well

Fuel bowl

Accelerator Pump System

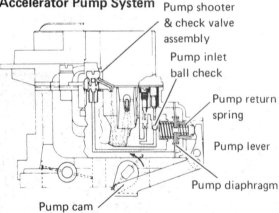

Pump shooter & check valve assembly

Pump inlet ball check

Pump return spring

Pump lever

Pump diaphragm

Pump cam

Choke System

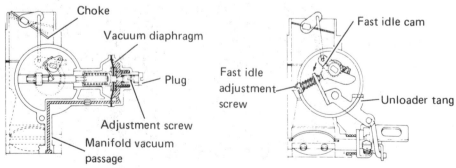

Choke

Vacuum diaphragm

Plug

Adjustment screw

Manifold vacuum passage

Fast idle cam

Fast idle adjustment screw

Unloader tang

Excess fuel vapor generated in the pump chamber during hot soaks is vented into the fuel bowl through a restriction.

Choke System—The automatic choke assembly on the carburetor body has a bi-metal thermostatic coil which winds up when cold and unwinds when hot. The coil is attached to a shaft which has a fast-idle cam and a linkage to open/close the choke plates. Engine coolant flowing through the water cover on the choke heats the bimetal coil to open the choke as engine temperature increases to operating conditions.

The choke system is brought into operation by depressing the accelerator pedal before a cold start, which allows the bimetal to close the choke plates.

A vacuum diaphragm and spring provide vacuum qualification of the choke opening once the engine is started.

Deceleration System—This system supplies additional mixture to the engine to ensure against misfiring during deceleration because this would increase hydrocarbon emissions. This system operates only at manifold vacuums higher than normally experienced at idle. When the carburetor is being used on an engine without a deceleration valve, the 1/4-inch-diameter brass tube on the air-horn/cover assembly should be plugged. The 5210 does not have the deceleration system.

During deceleration, high manifold vacuum is applied to a fitting on the carburetor air horn through a deceleration valve and vacuum line which are not part of the carburetor. This vacuum pulls fuel from a deceleration well in the fuel bowl, past a restriction. The fuel is mixed with air from an air bleed and the mixture is metered through a restriction into the cross passage connected to manifold vacuum. Another air restriction in this cross passage adds a larger quantity of air to the mixture as it is drawn through the vacuum line and deceleration valve into the intake manifold.

Accelerator pump in middle position 2. Moving pivot pin to hole 1 decreases pump capacity, outside position 3 causes an increase. Early models had only two holes. Pump-lever housing should be removed when changing positions. Support pump casting while removing and inserting pivot pin to prevent casting breakage. Pivot pin (arrow) is removed by driving with a small drift punch from the non-knurled side. Replace by driving from the knurled side.

HOLLEY 5200 & 5210 JET CHART

GREEN MAIN JETS	BRASS-COLOR MAIN JETS	HIGH-SPEED AIR BLEEDS	IDLE JETS
22BP-130A-227	22BP-103-130	22BP-110A-45	22BP-106-150
22BP-130A-239	22BP-103-132	22BP-110A-50	22BP-106-160
22BP-130A-243	22BP-103-133	22BP-110A-55	22BP-106-170
22BP-130A-255	22BP-103-135	22BP-110A-60	22BP-106-175
22BP-130A-263	22BP-103-137	22BP-110A-65	22BP-106-180
22BP-130A-275	22BP-103-140	22BP-110A-70	22BP-106-185
22BP-130A-283	22BP-103-142	22BP-110A-80	22BP-106-190
22BP-130A-299	22BP-103-145		
22BP-130A-311	22BP-103-147		
22BP-130A-325	22BP-103-150		
22BP-130A-346	22BP-103-155		
22BP-130A-357	22BP-103-157		
22BP-130A-374	22BP-103-160		
22BP-130A-404	22BP-103-165		
22BP-130A-423	22BP-103-170		
22BP-130A-455	22BP-103-175		
22BP-130A-477	22BP-103-180		
22BP-130A-524	22BP-103-185		

Brass-color jet sizes are in millimeters: -130 is 1.3 mm.
These sizes are typical, others may become available as the need develops.
There may be other green jets, but these are equivalents.

Two kinds of 5200/5210 main jets. Brass ones on left were used through 1975, are marked for diameter. Green-dyed ones on right are marked for flow. See text.

MAIN JETS: Two kinds!

In 1975 Holley began using a new series of main jets in Model 5200 and 5210 carburetors. These jets are sized and marked according to average flow. They are dyed green to avoid confusion with the formerly-used brass-color jets which are marked to indicate approximate diameter of the opening. Example: A 22R-130A-325 (marked 325) flows approximately 325 cubic centimeters per minute with a supply head of 50 centimeters.

The accompanying chart shows the relation of flow-rated jets to the diameter-marked jets.

5200 features include booster venturis 1, accelerator-pump shooters 2 (only primary side is drilled but secondary can be opened up), main air bleeds 3 with emulsion tubes 4 underneath. Idle air bleeds 5, idle jets 6 are in individual holders 7. Secondary system power well is 8, deceleration system well is 9. Power valve 10 screws into bottom of bowl. Main jets are 11.

Model 2110

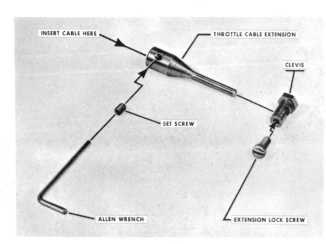

Model 2110 for VW's has an electric choke. Terminal 1 comes with an attached ground wire connected to a bowl screw. 2 connects to the choke lead from the primary circuit of the ignition coil. When used with a 12-volt system, wire indicated by arrows must be clipped off or the plastic choke housing will be "fried." Throttle linkage return spring hooks onto convenient grooved screw attaching choke housing. Fuel-inlet fitting works with stock VW hose.

The "Bug Spray" version of the Model 2110 is designed for Volkswagen engines. Two sizes are offered: 200 and 300 CFM. Holley recommends that you use these on an isolated-type manifold which connects one barrel of the carburetor to one side of the engine and the other barrel to the other side.

The 200-CFM version, List 6244-1, is the best choice for any VW for overall drivability. Its smaller capacity causes only a minor loss of top-end performance and low-speed response is much better than that of the larger version. List 4691-1, a 300 CFM unit, should only be used on 1800 cc and larger modified engines and then only on engines in very good condition. If the engine is "tired," an off-idle bog will inevitably result.

These carburetors are equipped with a a ported vacuum signal for 1973 and later vehicles. When installed on pre-1973 vehicles that use a pressure distributor as original equipment, the distributor must be replaced with a mechanical advance variety. The Bosch truck transporter distributor 231 129 019 (VW 211 905 205F) is usually used for VW's to be driven on the street and highway. More details on the correct distributor choice are provided in H. P. Books' *How to Hotrod Volkswagen Engines.*

Holley recommends and supplies intake manifolds to use with these carburetors. These manifolds have been designed to provide exhaust heat right up to the base of the carburetor. This has been done to preserve drivability and to ensure against icing in cold climates.

Throttle linkage assembly details. The loose parts are supplied in an envelope in the carton containing the carburetor. Entire linkage is designed for simple installation onto stock VW throttle cable.

BRIEF SPECIFICATIONS

This two-barrel carburetor flows either 200 or 300 CFM at 3.0-inch Hg pressure drop, depending on the venturi size. Throttle bores are 1-7/16-inch diameter in an SAE 1-1/4-inch two-barrel flange with a 1.87 X 3.24-inch stud pattern. The air-cleaner flange is 2-5/8-inch diameter.

Three castings are used to construct the carburetor: A zinc air horn and main body and a cast-iron throttle base. Removable booster-venturis are also zinc die castings.

A positive-drive accelerator pump is used. There is no power valve. The center-hung float is made of brass. The throttle linkage is designed for standard VW pull-type cable actuation.

An electric choke clip-connects to a primary coil lead. It works with either 6V or 12V systems. A wire lead is cut off for 12V use. The fuel-inlet fitting is designed for the VW fuel hose.

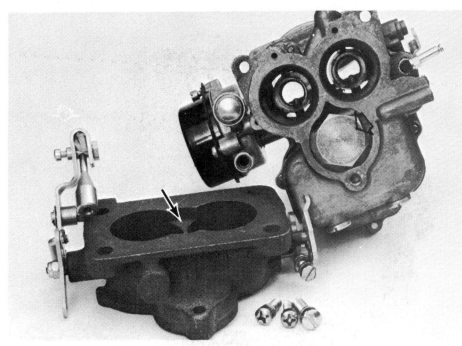

Manifold side of base shows plenum slot (arrow) which softens wide-open-throttle pulsing. Carburetor must be used on an isolated-tube manifold to avoid loss of mid-range response and drivability. Because this version of the Model 2110 is not equipped for a power valve, a plug is used in the main casting (outline arrow). This is a 200-CFM-er.

Float setting for either 200 or 300 CFM Model 2110 Bug Spray carburetor is checked with 3/16-inch drill. Drill on flat gasket surface of inverted bowl cover should just touch top of float.

Holley VW manifold for dual-port heads has exhaust heat on the runners and to the carburetor base to improve drivability and ensure against icing.

Below: Boost venturis called "nozzle bars" include idle air bleed 1, idle-feed restriction 2, idle-fuel passage around booster venturi 3 across to idle down leg on opposite side 4. Nozzle bars also contain main air bleed 5, emulsion tube 6 and the discharge nozzle (not shown).

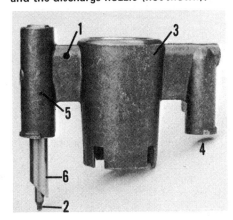

A set of wire drills and a pin vise are all you need to open up the idle-feed restrictions if your engine has a flat spot off the idle. Enlarge the restrictions 0.002 inch at a time until you have the problem cured. If you go too far, the tip of the restriction can be soldered closed and re-drilled to the smaller size required.

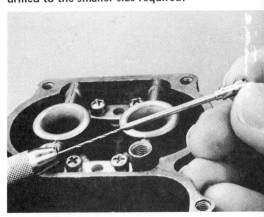

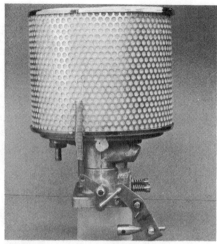

Two views of the Buggy Spray show air cleaner removed at left—installed at right. The air cleaner reduces air flow capacity a mere 5 CFM from the 300 CFM rating. Cleaner element should be sealed to carburetor flange and to cleaner top with Silastic RTV or a similar sealant to keep dust from entering around the ends of the cleaner element. The cleaner—called "Bug Catcher" is 80R-26AS. For really rough terrain, some off-roaders add tubes to the bowl-cover vents (under large washer) and extend these to the top of the air cleaner. More details on off-road air cleaners are in H. P. Books' *Baja-Prepping VW Sedans & Dune Buggies*. Carburetor may have to be used with a mechanical-advance distributor as described in text.

Slot between throttle bores (arrow) base provides a small-volume plenum to soften wide-open-throttle pulsing. Idle screw removed to show seal.

Model 1901

The 1901 "Buggy Spray" is designed for off-road use on VW sedans or dune buggies equipped with VW or Porsche engines of at least 1300cc or larger displacement. It should be used on an isolated-tube manifold which connects one carburetor barrel to each side of the engine. The carburetor's excellent angularity characteristics meet military specifications. The List No. is 6113-1.

BRIEF SPECIFICATIONS

This two-barrel carburetor flows 300 CFM at 3.0-inch Hg pressure drop. Throttle bores are 1-7/16-inch diameter in an SAE 1-1/4-inch two-barrel flange with a 1.87 X 3.24-inch stud pattern. The 5.0-inch-diameter air-cleaner flange allows a tall air cleaner to enclose the concentric center-mounted fuel bowl.

Three major castings are used in constructing the carburetor: A zinc fuel bowl and cover and an aluminum throttle body including the venturis. A booster-venturi casting is sandwiched between the fuel bowl and the throttle body.

A positive-drive accelerator pump is used. The inverted-type power valve provides correct F/A ratios, even though the carburetor is subjected to strong pulsing at wide-open throttle. High-load full-throttle enrichment is provided by pulsing which pulls fuel from the discharge nozzle in both directions of air flow. At high manifold vacuums and part-throttle, the power valve is open. As full-throttle, and low manifold vacuums (below 3 inches Hg) are approached, the power valve closes so only the main jet feeds fuel to the main system.

The center-hung brass float has a bumper spring and the inlet needle is also spring-loaded. Dirt/dust seals are provided on the front-facing idle-adjustment screws, on the throttle shaft and on accelerator-pump linkage. Thus, when an air cleaner is installed and sealed to the flange and cover, the carburetor becomes dust-proof. The throttle linkage is designed for standard VW pull-type cable actuation. The fuel-inlet fitting is designed for use with the stock VW fuel hose.

On pre-1973 vehicles that use a pressure distributor as original equipment, the distributor must be replaced with a mechanical advance variety. A Bosch truck/

transporter 231 129 019 (VW 211 905 205F) is usually used for the replacement. More details on the correct distributor choice are provided in H. P. Books' *How to Hotrod Volkswagen Engines.*

This carburetor does not have a choke system.

DESCRIPTION OF OPERATION

With the exception of the inverted power valve—pictured in the power system section of Chapter 2—the operation of the carburetor is essentially as discussed in the earlier section on how your carburetor works.

Many important features are illustrated in this photo. Spring-loaded needle & seat 1, inverted power 2 (closes when manifold vacuum drops to 3 inches Hg), main jets 3, main air bleed 4 in top of emulsion tubes, idle-feed restrictions 5 (idle air bleeds are in bowl cover), and accelerator-pump discharge check 6. Accelerator-pump linkage is sealed when it enters bowl area at 7. Float-bumper spring is 8.

1901 fuel level is easily wet-checked by removing float-bowl cover. Run engine for a moment to establish the fuel level, then shut if off so carburetor will not be shaking. Fuel surface should be 17/32 inch below the gasket surface. Here a plastic ruler has been cut to make a gage for making the measurement. Dry checking is not recommended because inverting the carburetor allows the pump check and idle tubes to fall out.

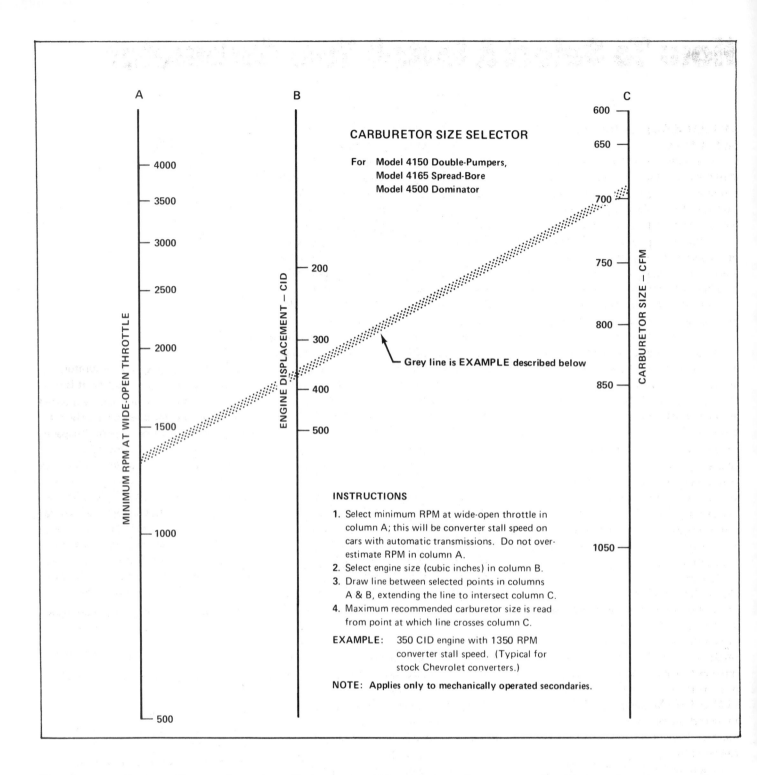

CARBURETOR SIZE SELECTOR

For Model 4150 Double-Pumpers,
Model 4165 Spread-Bore
Model 4500 Dominator

A — MINIMUM RPM AT WIDE-OPEN THROTTLE

B — ENGINE DISPLACEMENT – CID

C — CARBURETOR SIZE – CFM

Grey line is EXAMPLE described below

INSTRUCTIONS

1. Select minimum RPM at wide-open throttle in column A; this will be converter stall speed on cars with automatic transmissions. Do not over-estimate RPM in column A.
2. Select engine size (cubic inches) in column B.
3. Draw line between selected points in columns A & B, extending the line to intersect column C.
4. Maximum recommended carburetor size is read from point at which line crosses column C.

EXAMPLE: 350 CID engine with 1350 RPM converter stall speed. (Typical for stock Chevrolet converters.)

NOTE: Applies only to mechanically operated secondaries.

More about using the chart — If your car has an automatic-transmission, make sure you know the converter stall speed before using the chart. If in doubt, use the figure shown for a typical Chevrolet converter (1350 RPM). If you are using a modified converter for a racing application, make sure the stall speed is what you think it is.

If your car has a manual transmission, use the lowest RPM at which you use wide-open throttle. This must be a very conservative RPM (on the low-RPM side, that is!) and should be found by observing your own driving habits in the vehicle involved. Watch your tachometer!

The heavier the vehicle and the lower the numerical axle ratio (higher gear ratio) — the lower this RPM must be.

With engines from 300 to 400 CID, the right choice usually works out to be a 650 to 700 CFM carburetor. A light car, such as a Camaro, Mustang or Duster may be able to use a 700 or 750 CFM unit, especially with a high numerical gear ratio (low gear ratio).

When in doubt, select a smaller carburetor size because it will typically give better acceleration times — even though power may fall off slightly at top RPM. You can believe that you'll be hap-

pier with the smaller carburetor nearly every time!

Regardless of all the evidence to the contrary, a lot of carburetor buyers "psych" themselves into believing that bigger is better. Thus, Holley sells more large carburetors (800 and 850 CFM) because of the widespread fallacy that if a 650 is good — an 850 must be just that much better. This is not true!

Co-author Urich regularly gives this advice in relation to air-flow capacity, **"Don't buy it if you can't use it!"**

How To Select & Install Your Carburetor

MECHANICAL SECONDARY SELECTION

A carburetor with mechanical secondaries has an inherent advantage over any carburetor of the same capacity with a controlled secondary (air valve or diaphragm-operated). At any engine speeds lower than the speed where the controlled secondaries or air valve are completely open, the mechanical-secondary carburetor will have a lower pressure drop across it at wide-open throttle. Because the manifold vacuum is lower, the charge will be denser. . .resulting in greater engine output.

We must use greater care in selecting the correct size mechanical-secondary carburetor. When we slam the throttle wide open on a controlled-secondary carburetor, it operates only on the primary side until a specified higher air flow is reached. Thus, primary velocities are high, giving a strong metering signal. When we punch a mechanical-secondary carburetor wide open, air flows through all of the venturis, hence velocities and signals are low. Double-pumpers depend on the pump shot to supply correct mixture until adequate air flow is established to start the main system.

The larger the carburetor—the higher the air flow must be before the main systems begin to feed. If the carburetor is too large, the pump shot will be consumed before the main systems start. This will result in a sag or bog. This is why double-pumper four-barrel carburetors are offered in so many sizes: From 600 CFM to 850 CFM in 50-CFM increments. When using these carburetors, greater care must be taken to match the carburetor with the engine/vehicle package.

Co-author Urich says Holley Technical Representatives constantly see too-large carburetors installed. So he put together this handy chart to aid in selecting double-pumper type carburetors. Follow it and you'll get the right one for your application.

IMPROVED BREATHING INCREASES VOLUMETRIC EFFICIENCY AND POWER

The stock passenger-car engine is a reasonably efficient air pump—over a fairly wide RPM range—from low RPM to 5,000 RPM or slightly higher. It provides a reasonably flat torque curve over this operating band. Its pumping capabilities can be idealized (optimized) to provide better pumping within a narrower RPM band by improving breathing. By reducing restrictions which cause pressure drop, charge density reaching the cylinder is increased. Inprovements are typically made by changes to the carburetor (higher capacity), intake tract (manifold through the ports and valves), exhaust system (headers or free-breathing mufflers) and valve timing (camshaft). Such changes are similar to the things that a pump designer would do to make an efficient pump—he'd work to optimize the performance *at a particular RPM*. But, there is a trade-off. When you make the engine a better pump, the torque peak and the entire torque curve are lifted to a higher RPM band. Lower RPM performance is *reduced* accordingly.

Not all engines can be improved for higher performance—better breathing—without extensive modifications. This is because the designers have purposely optimized the low-RPM performance with measures such as small-venturied carburetors, tiny intake manifold passages and ports, small valves actuated by short-duration camshafts with lazy action, low compression, combustion-chamber design and restricted exhaust systems. Truck engines, low-performance passenger car engines and industrial engines are good examples. Keep such factors in mind when you are selecting a carburetor, because it is difficult—if not impossible—to upgrade the performance of an engine with such built-in restrictions. As they say, "You can't make a race horse out of a mule." If the engine is so restricted that it produces peak power at 4,000 RPM, selecting a carburetor on the basis of feeding that same engine at 6,000 to 7,000 RPM is not wise. It's unlikely that the engine will ever run at those speeds—at least, not without extensive modifications. And, a too-large carburetor will definitely worsen the performance which was previously available.

NOTE: When comparing carburetors on the basis of their flow capacity, it is important to know that one- and two-barrel carburetors are measured on a different basis than four-barrels. The flow capacity of one- and two-barrel carburetors is measured at a pressure drop of 3.0 inches Hg. This figure has been used for many years because this is a typical wide-open-throttle pressure drop across the carburetor (manifold vacuum) in low- and medium-performance engines. However, the four-barrel carburetors are measured at a pressure drop of 1.5 inches Hg. To relate the two measurements, use the formula

$$\frac{\text{CFM flow at 3 in. Hg}}{1.414} = \frac{\text{Equivalent flow}}{\text{at 1.5 in. Hg}}$$

EXAMPLE: For a 500-CFM two-barrel

$$\frac{500}{1.414} = 354 \text{ CFM four-barrel equivalent}$$

Some automobile manufacturers rate Holley carburetors at other pressure drops, giving different flow ratings than Holley's specification. For instance, Chrysler rates the 2210 two-barrel at 2.0-inches Hg pressure drop at WOT and uses a reduced CFM rating. Flow rates shown in Holley catalogs are standardized at 1.5 in. Hg pressure drop for three- and four-barrel carburetors; 3.0 in. Hg for one- and two-barrel carburetors.

Flow rates shown in Holley catalogs are standardized at 1.5 in. Hg pressure drop for three- and four-barrel carburetors; 3.0 in. Hg for one- and two-barrel carburetors.

UNPACKING YOUR CARBURETOR

If you are buying your carburetor in a store, unpack it before you pay for it. Look at the *inside* of the box as you take out each piece. Note whether one side of the box shows evidence of being damaged from movement of the carburetor. If such is obvious, check that side of the carb very carefully. Look it over slowly and thoroughly to make sure that there has not been any shipping damage. Don't make a hurried examination. Although Holley packaging engineers constantly improve the packaging, their best efforts are often in vain when the freight companies mishandle the shipments. You should be aware that it is quite possible for the *outside* of a carton to look perfectly intact—yet the contents may be damaged.

Give the carburetor a visual inspection, then hold the carburetor in your hand as you operate the throttle lever to make sure that the throttles open fully and close without binding or sticking. A bent throttle lever can occur. When it does, it is possible for the throttles to remain partly open (not return fully to idle or fast-idle position) when the lever is actuated.

If you have ordered your carburetor through the mail, give it the same inspeciton immediately upon its arrival. Don't just open the box and look inside to make sure that it is the correct carburetor—although that *is* important, as sometimes the wrong carburetor may be shipped—check it over thoroughly as just mentioned.

Should there be a problem, proceed carefully if you try to fix it yourself. If it is a simple thing—such as a bent throttle lever—you may be able to remedy the problem yourself easier than returning the unit. However, if the casting has been damaged, it is obvious the part will have to be replaced.

The firm you bought the carburetor from will replace it. If you ordered the carburetor from out of town, find out what the supplier requires *before you send it back for replacement.* In some instances, the settlement of damage claims will be between yourself and the freight company (called *the carrier*). In other cases, the supplier will take care of this for you.

Order the carburetor *early.* It may not be obvious to you, but one of the first things to remember in building your engine is to get the correct parts together and then get the car assembled and tested before setting impossible dates for completion and first competition attempts. Many cars do not show up at major events because the builder has been over-optimistic about the amount of time required to obtain all of the parts and then to assemble and test the combination. The owner may lose an expensive entry fee as well as missing the race.

In the case of a Holley carburetor, it is possible the supplier may not have the unit you need in stock. He could have just sold the last one and be waiting for a new shipment to arrive. Or, the unit you require may be an extremely popular one which has been selling so well that the factory is temporarily unable to keep the distribution channels filled with enough carburetors.

If you ordered your carburetor early in the game—ahead of the time when you absolutely had to have it, then this problem will probably not give you any trouble. But, if you wait until the day before you have to run your machine, it is conceivable that you could end up buying a carburetor which is not exactly the one that you want or need. This usually means you will buy the right carburetor later on—or else keep on using a carburetor not precisely right for your application. Plan ahead and do your carburetor shopping early. You'll be glad that you did. Be sure to get the other pieces you need at the same time, including fuel lines, extra gaskets and jets, air cleaner, tubing wrench, new fittings and a fuel filter.

BEFORE INSTALLING YOUR CARBURETOR

Look the carburetor over carefully when you take it out of the box. Make sure that the carburetor operates easily, opens to full throttle on both primary and secondary barrels (assuming you have a two-barrel progressive carburetor or a four-barrel carburetor), and returns easily to idle position. The throttle shafts are equipped with adjustable stops. Throttle/s should return to rest against the adjustment screws when the choke is open.

Check for loose screws which attach the top of the carburetor (if it has a top cover), the screws which attach the float bowls (if it has bowls), and the screws which hold the accelerator-pump shooter/s in place—if these are used on the carb that you have purchased. Also check the screws which hold the castings together. Gaskets will compress after they have been installed and it is important to check the screws as described so that there will not be any fuel or air leakage after the carburetor is installed.

If you happen to take the carb apart to look at its insides, put it back together immediately upon finishing your inspection so that the gaskets will not shrink into an unusable condition. And, be sure you check the accelerator pump/s for correct clearance at the end of the stroke if you are working with a two- or four-barrel unit with the diaphragm-type pumps.

A lot of magazine articles have indicated it is a good idea to take the throttle butterflies off of the shafts or to loosen the screws so that the throttles can be centered in the throttle bores. DON'T DO IT. Throttles are set against adjustment stops and checked for the correct air flow before they leave the factory. If you hold the carb up to a light, you will see clearances around the throttles . . . that's the way it is *supposed* to be! You can end up creating more trouble for yourself than you can imagine if you happen to break off one of the brass screws. These are staked at the factory so they will not turn. A quick visual inspection of the carburetor will indicate which screws have been staked.

In general, you should plan to bolt the carburetor on and run it before making any changes. Forget all the rumors and bench-racing discussion that you have been participating in lately. Start with what the Holley engineers have found to work successfully. They produce hundreds of thousands of carburetors every year—for all kinds of applications—and chances are awfully good that the carburetor will be very close to correct when it is bolted onto your engine.

Where the pump/s are in removable fuel bowl/s, check for correct accelerator-pump adjustment. Open the throttle wide, then actuate the diaphragm lever by hand. This is the 2-1/8-inch-long lever which pivots in the pump cover casting. This lever must have an additional 0.015—0.020-inch travel available at wide-open throttle to ensure the pump is not bottomed in its housing. This also ensures the throttle will not stick in the open position because of excessive friction of the pump-operating lever against the plastic cam. Also check that the pump-operating lever contacts the plastic cam at one end and the diaphragm lever at the other. While you should be able to wiggle the diaphragm lever back and forth slightly (horizontally) with your fingers with the throttle at idle, the slightest movement of the throttle should be transmitted through the pump operating lever, causing the diaphragm lever to move.

The only exception is that when a green cam is used, 0.010 clearance is used at WOT to get correct actuation at idle.

INSTALLING YOUR CARBURETOR

1. Remove the air cleaner, carefully detaching any lines to the cleaner and marking them with masking tape so they can be reassembled correctly.
2. Remove existing carburetor as follows (not all items apply to every application, of course):

A. Remove steel fuel line fitting carefully because it is reused in most instances. A fuel-line wrench which contacts four of the nut flats should be used because the fuel-line nuts are often quite soft and tend to round off instead of turning. When this happens you will probably have to resort to the handy Vise-Grip pliers to get the nut out. This will totally wreck the nut. Then cut off the fuel line very close to the end of the tubing. Going through the drill of cutting and reflaring requires a tubing cutter and a flaring tool. These may be rented from your nearby auto-parts house in some instances. Getting the line off the engine to allow this work will require disconnecting the fuel line at the fuel pump. If the nut at the carburetor rounded off, you can be sure the one on the fuel pump will also. So go buy a tubing wrench before taking the one off at the fuel pump or you may have to create an entire new fuel line to get the carburetor installed. The wrench is cheaper in the long run.

B. Remove throttle linkage and automatic-transmission controls from throttle lever. Take off throttle-return spring and note its anchor points for correct reassembly.

Essential pre-installation checks for each accelerator pump on carburetors with removable fuel bowls: (1) no clearance between the actuating-lever screw and the diaphragm lever at curb idle, (2) you should be able to move the diaphragm lever enough farther at WOT to allow inserting a 0.015 to 0.020-inch feeler gage as shown. Text provides more details.

Throttle and transmission linkages, smog-equipment connections and so forth can seem hopelessly complex if you forget where each piece goes. Make a sketch BEFORE you take your manifold and/or carburetor off. Tag the various lines and connections with tape "flags" to identify each plainly. If you have a Polaroid camera, take a picture of each side of the engine after removing the air cleaner and tagging any lines that lead to it. Or, make a drawing to aid your memory.

Holley adapters allow mating carburetor flanges to different manifold stud/bore patterns. Although such adapters must be used in some instances, it is preferable to use a manifold with the matching stud/bore pattern for the carburetor to avoid flow losses. In the case of the Holley 4500, the flow loss is severe when it is adapted to a non-4500 manifold.

C. Remove PCV hose if it is attached to carburetor.

D. Remove distributor spark hose/s, labeling these to keep vacuum and timed-spark hoses separate for correct reinstallation.

E. Remove fresh-air hose if one is used and label it. Remove any other hoses such as gulp-valve hose (for air-injection-equipped cars) and connections to power brakes, etc. Label these.

F. Detach choke linkage at the carburetor, noting whether it pulls or pushes to close the choke plate.

G. Remove the nuts and/or bolts and lockwashers attaching carburetor to manifold, taking care to avoid dropping into the carburetor. Covering the top of the carburetor is a good idea at this point. If you drop any nut, lockwasher, cotter key, piece of linkage or whatever, STOP. Find and recover it *before* proceeding. Observe and note the position of any brackets held on the engine by the carburetor-attachment hardware. Lay these aside so they will not fall into the manifold when the carburetor is pulled off. Before you lift the carburetor off of the manifold, check carefully to make sure nothing loose can fall into the manifold . . . such as a nut, bolt, fitting, cotter key or whatever.

H. Carefully pull the carburetor off the manifold. If it sticks, tap it gently at each side with a rubber or plastic mallet. Clean the intake-manifold mounting surface carefully, taking care to keep gasket pieces from getting into the manifold.

3. Install new carburetor as follows (not all steps apply to every application):

A. Install any new studs which are required. Use double-nut technique to install studs. Use Loctite on studs if available. Install manifold-flange gasket. If a metal heat shield is used to protect the carburetor base from exhaust gases in the heat-riser passage, place the thick (0.080-inch) manifold-flange gasket under steel heat shield. No gasket is used between the shield and the carburetor. Pay attention to the gasket combination used by the automobile manufacturer. Duplicate it as nearly as possible.

B. Some items used on the original carburetor may have to be transferred to the new one. These may include linkage ball connections, or lever extensions. If the original throttle lever has a non-removable ball, a corresponding ball will be found in the loose parts supplied with the carburetor. These are in a small envelope in the carburetor box.

C. Some items on the new carburetor may have to be removed to adapt it to a particular application. Examples include a lower throttle-lever extension on some 4165 carburetors. This is removed on some big-block applications. The 500 and 650 CFM two-barrel 2300 and some 4150 carburetors with universal throttle linkage may require removing a Chrysler-type lever to adapt the carburetor to Chevrolet and Ford installations. Details on the universal linkage just mentioned are contained in the section on the 3310 carburetor elsewhere in this book.

D. One of the easiest ways to prevent manifold leaks is to check the carburetor base gasket. Hold it against the manifold flange to make sure there is plenty of clamping area to hold the gasket in position. Look to see whether any openings do not match up. When you are satisfied the gasket is correct for the manifold, hold the gasket against the carburetor base to see if it fits the throttle bores, mounting holes and general shape of the base. Place gasket on manifold.

E. Place carburetor on manifold and add any brackets held in place by the attachment hardware—including throttle-linkage and solenoid brackets if such are used. Install nuts and/or bolts and tighten them gently against the carburetor flange. Then cross-tighten the bolts/nuts alternately to 5 to 7 lbs. ft. torque. Overtightening can cause a warped throttle body which will bind the throttle shafts so throttle action is stiff. Vacuum-operated secondary throttles may never open if the base is warped. In the worst case, overtightening may snap off a corner of the carburetor base. That *is* expensive and time-consuming.

If the carburetor has vacuum-operated secondaries, hold the throttle wide open and operate the secondaries by hand, making sure they will open easily against the spring and are closed freely by spring action. Any binding should cause you to suspect uneven tightening of the nuts holding the carburetor to the manifold. If they are overtightened or tightened unevenly, the base may warp, causing binding of the secondary shaft. If you fail to check this point, the secondaries may never work and you'll lose a lot of performance.

F. Connect throttle and transmission linkage. Connect throttle-return spring. Operate linkage to ensure nothing binds as the throttle is fully opened and then closed. Make sure that the throttle returns all the way to idle without sticking or binding with choke held wide open.

Make sure the levers don't hit anything as the throttle linkage is actuated. Make this check by having a friend actuate the linkage with the foot pedal as you observe the operation of the linkage. Look in the carburetor to see for yourself whether the throttle plates are 100% open in both the primary and secondary barrels. Don't try to make this check by operating the throttle with your hand at the carburetor or by looking at what the levers *appear* to be doing. Slack or play in the linkage may allow full throttle when moved at the carburetor, but give less than full throttle when the foot pedal is depressed. If the foot pedal movement does not give full throttle, examine the linkage to see what minor changes or adjustments may be needed in the linkage connecting to the carburetor. Do not change the linkage or levers on the carburetor itself because simpler adjustments are usually provided on the vehicle cable housing or linkage rods.

Don't think checking for full throttle is only for amateurs. Experts have been tripped up on this point . . . whether they care to admit it or not.

Check both the pump and the throttle levers to make sure nothing is in the way. Sometimes the carburetor has to install on an aluminum spacer to get clearance for high-capacity accelerator pump/s. Install air cleaner and make sure it does not hit any portion of the linkage. Operate the hand choke if there is one so you can determine whether any clearance problem exists.

G. Install choke linkage for divorced choke. Hold throttle lever partially open and operate choke manually to make sure it operates freely.

H. Reconnect appropriate hoses to carburetor. If additional fittings should be required, transfer them from the old carburetor if you can.

I. Flush fuel line. Pull primary wire from coil and insulate its end with a piece of tape so it won't spark. Hold a can under the open end of the fuel line and catch the fuel as the starter is used to crank the engine. Connect fuel line to carburetor or fuel-line assembly. In some instances a minor bend may be required in the fuel line or you may have to shorten the fuel line and reflare the end.

J. Remove air-cleaner stud from old carburetor and install it in the new one, or use one provided.

4. Crank the engine or turn on the electric fuel pump to fill the carburetor with fuel. Check the fuel-inlet fitting/s for leaks and tighten fittings to eliminate any leaks.

5. Start the engine. Depress the accelerator pedal to floor and allow to return to normal. This charges the manifold with fuel, sets the choke and fast idle. With foot off accelerator pedal, crank engine until it fires, repeating the previous two steps as required.

A. If the carburetor floods over, there may be dirt in the needle and seat assembly.

B. If flooding continues, stop to correct cause. First try tapping the carburetor lightly with the handle of a screwdriver. If this does not stop the flooding, remove the fuel line and take off the carburetor top or float bowl (depending on how that particular carburetor is constructed). Check the float setting. If it is o.k., the chances are there is dirt under the needle. If the needle and seat are separate items, pull the needle out of the seat and blow out the seat with compressed air from both sides. If the needle and seat assembly is not of the take-apart type, unscrew the entire needle and seat from the float bowl and agitate the assembly in solvent, then blow it out with compressed air.

6. Fast-idle setting is usually adequate for most vehicles. If a particular application requires adjustment, use the following procedure: With engine running, advance throttle and place fast idle cam so speed screw contacts top step of cam. On this step, set fast idle at approximately 1600 RPM.

If your vehicle has T.C.S. (Transmission Controlled Spark advance) your adjustment of fast-idle speed is being made with no vacuum advance in neutral. When the engine is cold, there will be vacuum advance and the speed may be too fast (objectionable). Make subtle changes to get it where it suits you.

GET ACQUAINTED WITH IT!

Once you have installed your new carburetor, live with the new combination for a few days before beginning your tuning efforts. Naturally, you will want to set the idle correctly (to exact specifications if your car has emission-control devices), check the ignition timing with a timing light and make sure the spark plugs, points, cap and wires are all in tip-top shape.

By driving the car, you will learn what you have. If you are drag racing, get your times consistent by working on your driving technique *before* you start tuning. There are enough variables without adding inconsistent driving. This is especially important if you have changed from an air-valve-secondary carburetor (such as a Q-Jet) or a vacuum-operated-secondary carburetor to a double-pumper with a mechanical-secondary linkage.

Many details related to tuning are provided in a later chapter on the carburetor and performance.

MAC S-141 1-inch end wrench (extra-slim) is ideal for fuel-inlet nuts. Works on the diaphragm-type power valves, too—if used with care.

Avoid gasket stacks or packs—These compress unevenly so that the carburetor base warps to bind the throttle shaft. They often allow one of the corners of the throttle base (mounting flange) to break off. Always use a thin gasket such as the 0.025-inch thick one supplied by Holley with the diaphragm-secondary carburetor. Otherwise use the gasket supplied with the carburetor. No gasket sealer is required if the carburetor base and manifold flange are clean and flat. Do not stack gaskets together to get clearance for lever operation, regardless of how badly you want to get your engine running. Do it right or you could end up buying an entire new carburetor body or throttle base. Because these parts are not ordinarily stocked by even the largest dealers, there could be a long wait for pieces to reassemble your carburetor. If the throttle body is not a separate part, you will have to buy a new carburetor because the main body is not sold as a replacement part. If you need additional clearance to install a carburetor with a high-capacity accelerator pump, make a 1/4-inch-thick aluminum spacer and use a thin gasket on top of the spacer and one on the bottom.

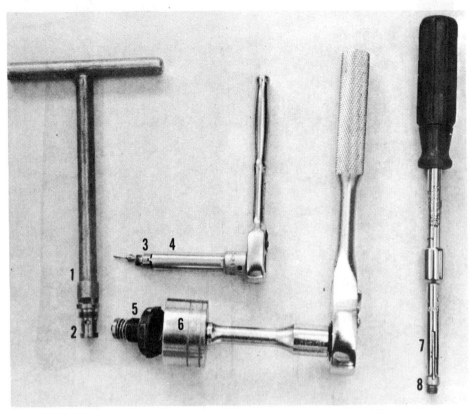

Although special tools for Holleys were originally made by several tool firms, the only supplier now is Kent-Moore Tools, 1501 S. Jackson St., Jackson, MI 49203. Tool 1 in photo is hand-made X-driver with T-handle for piston-actuated power valve 2. Piston-actuated power valve 3 can be removed or installed with special socket 4 (use K-M J-10185) which is a screwdriver with a hole in the center. Diaphragm power valve 5 is used with socket 6 (K-M J-10234). Main-jet socket (K-M J-10174-01) is for main jet 8.

More tools: Bottom pliers are for removing/installing Corbin clamps used on fuel and water hoses. They are made by several firms. Two tubing wrenches in center are Sears Craftsman, Popular 1/2—9/16-inch size is not shown but should be part of your set. Top MAC S-141 wrench is needed for fuel-inlet nuts.

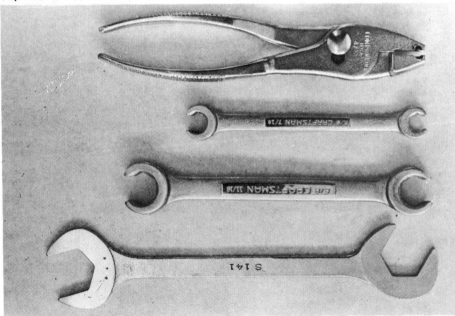

CHECKING FUEL LEVELS OR FLOAT SETTINGS

Installing the carburetor and pumping fuel to it is the only easy way to check whether the needle/seat will seal correctly. Turning the carburetor upside down and blowing into the fuel inlet is not an adequate test because your lungs cannot develop as much pressure as the fuel pump. When the carburetor is installed and the engine is started, observe whether there is any flooding into the manifold. If there is no seepage or flooding into the throttle bore, chances are the fuel level is o.k. If the carburetor does not have externally adjustable needle/seat assemblies and sight plugs, give the car a test drive. If the car can be driven normally without dying at stops, you can assume that the fuel level is probably o.k. If the fuel level is too high, fuel will spill into the throttle bore/s and the engine will die out at stops or after severe braking.

High-performance Holley carburetors have the advantage of externally adjustable needle/seat assemblies and fuel-level sight plugs which allow checking fuel levels and setting them without taking the carburetor apart and without taking it off of the engine.

If the engine has a mechanical fuel pump, lower the fuel level by loosening one bowl screw and allowing some fuel to leak out. Retighten the screw. Then run the engine to see whether the fuel level returns and stays where it is supposed to be. If you adjust the needle assembly so the level is just below the point where it would spill through the sight plug hole and check to see it stays there, then you can raise the level until the fuel just spills through the plug hole. That way, you have assurance the needle and seat assembly is holding and the level is as it should be.

Any installation with an electrical pump can be checked by turning the pump on without running the engine to make sure that the needle and seat assemblies are holding and not "creeping." Then run the engine to make sure the level stays where it should. Be sure to reinstall the sight plugs after checking levels.

The main problems with fuel-level changes are caused by the forces imposed on the carburetor when it is shipped. Jarring forces encountered in transportation can bend the tabs which establish the float/needle relationship. We've even seen float bumper springs dislodged from the underside of the floats.

Once a carburetor is installed on a car, the fuel level established by the float setting seldom changes. This means the simple task of setting the level is usually a one-time thing with occasional checks to make sure the level is still o.k.

Don't raise or lower the fuel level except to get it to specification. It has been calibrated at the factory for cornering, turns, spin outs, brake stops and accelerations.

Float levels are most critical with high G-loadings because those affect fuel handling in the fuel bowl—Slosh problems in the fuel bowls are severe. Just imagine trying to hold a cup of coffee in a car without sloshing during a burn-out or a panic stop! That should give you some idea of what happens inside the carburetor fuel bowl/s. The only time that the float level might be changed would be for drags or super-speedway use. The secondary level might be raised slightly, but expect a brake-stop stall as the result.

Avoid taking off the bowl cover or carburetor top on any carburetor which uses a built-in vacuum-piston for the power valve—unless you have a replacement gasket handy. If you take the top off and the gasket tears, especially in the area where the vacuum passage from the carburetor body enters the mating passage which supplies vacuum to the power-valve piston, the power valve may be held open constantly. Where a carburetor is being assembled and disassembled for calibration purposes, it is common practice to apply talcum powder to both sides of the gasket so sticking will not occur.

BLUEPRINTING HOLLEY CARBURETORS

"Blueprinting" has become an important word in the enthusiast's vocabulary. The word, as applied to an engine, means bringing all of the parts of a production engine into the correct relationship by matching and mating, balancing and remachining as required to get the combination into a near-perfect state suitable for performance application. Blueprinting is often done on brand-new engines right out of the crate or just off of the show-room floor. This attention to detail can provide impressive performance improvement as an immediate payoff for the time and care invested in the project.

"Blueprinting," is now applied to the precise preparation and rebuilding of any component in the engine or chassis. But, it is a mistake to use the word or concept in relation to a Holley high-performance original-equipment or production-replacement carburetor. Tearing a new carburetor apart and attacking it with drills, trick linkages and so forth is like overhauling a new watch or micrometer when you first get it . . . something you probably wouldn't consider doing. There would be too much chance of ruining it—and voiding the guarantee—before you could use it.

Carburetors are made very precisely because they have to accomplish an extremely accurate job of metering fuel and air into the engine. *They have to be correct before they are shipped.* Every single carburetor which will be used as original or replacement equipment on a production automobile must pass critical computer-controlled tests which check whether that individual carburetor performs to tight specifications which ensure it will provide correct performance (including low emissions) when it is installed. If the carburetor is "out of spec," any "blueprinting" must be done by Holley because the unit has to pass the qualification tests before it can be shipped. These requirements are laid down by law.

Production techniques used to manufacture these carburetors have been developed over years of making carburetors to be right when an automobile manufacturer—or an automotive enthusiast bolts one onto an engine. Thus, regardless of what you may have read to the contrary, very little needs to be done—or can be done—to one of these precise devices to "make it better" before you use it. If "trick" things or special assembly techniques were required to make the carburetor right for high performance, these would be incorporated and fully tested before the design was released to one of the Holley manufacturing plants.

The authors are fully aware that numerous magazine articles are printed every year—pointing out "fixes" to be made to new or used carburetors for more performance. Almost without exception, these have been tried at Holley's engineering labs and in field tests—and discarded as being unworkable or not generally applicable. Great care in selecting, installing and tuning your carburetor is the best "trick" you can pull out of your bag or tool box to get the utmost in performance and satisfaction.

WHICH CARBURETOR IS THE RIGHT ONE FOR YOUR ENGINE?

There are enough different Holley carburetors available as aftermarket models that any engine can be correctly mated with a carburetor that is really RIGHT ON for the engine and application.

Some builders fall in love with a certain Holley size, such as the 850 CFM and disregard all of the other sizes, regardless of the fact that a larger or smaller carburetor might be correct for matching a particular application. The reason so many different types of carburetors are built is there are different kinds and sizes of engines and their requirements are not the same. Holley would certainly prefer to build fewer types of carburetors if this could be done and still match the various needs of the engines which will use them. Now, it is obvious from the outside that the carburetors—many of them, at least—are quite similar in many ways. But, the mechanical secondary, vacuum secondary, double pumper, single pumper, non-pull-over discharge nozzle with remote check valve, discharge nozzle with check valve just under it, rubber-type pump-inlet valve, ball pump-inlet valve, side-hung and center-hung float, and so on are all variations born because of specific needs. It may not be obvious if you are not a manufacturer, but every change costs money to design, to build—and then to supply as complete assemblies or as replacement parts. So, these pieces exist for a reason and after you have read this book, the reasons will be easier to understand—and you'll be right at home when you specify a particular carburetor for your own or a friend's engine.

Fuel Supply System—Read the last chapter for important details on fuel-lines, fuel filters and fuel pumps.

Adapters—A carburetor should be installed directly onto its manifold without an adapter whenever this is possible. Adapters do not usually provide optimum air-flow characteristics. Many actually reduce the carburetor's flow capacity. In general it is best to buy a carburetor that fits the manifold to be used. Or, it is sometimes possible to drill and tap the manifold to fit a slightly different stud pattern. In the case of a four-barrel manifold with "standard" pattern, the double-pumper or one of the vacuum secondary carburetors should be used. Or, if the manifold is designed for a spread-bore stud pattern, the 4165, 4175 or 4360 can be used. An adapter may raise the carburetor high enough so a special air cleaner will be required to gain hood clearance. And, in some instances throttle and choke linkage and fuel-line-attachment problems may also be caused.

BASIC WARRANTY POLICY

Holley Carburetor Division warrants each Holley product, or parts thereof, to be free from defects in workmanship and material if product or parts are properly installed, adjusted to specification as required and subjected to normal use and service.

Holley Carburetor Division's total obligation under this warranty is limited to repair or replacement of any Holley product or part found defective within 90 days after installation.

Conditions Not Covered by Warranty—Failure caused by the following conditions voids warranty on Holley carburetors: Poor quality of gasoline, dirt or other comtaminants, gums or varnish, water or corrosion, modification, improper fuel-inlet pressure, installation damage, improper adjustment, improper installation, faulty repair.

USE THE RIGHT GASKET—Holley supplies the correct gasket with the carburetor when you buy it. When replacing a gasket, always try to duplicate the one which came with the carburetor: Shape, hole sizes, bolt pattern *and* thickness. Gasket thickness can be very important because it often relates to the choke rod for attachment to a divorced choke bi-metal. The wrong gasket thickness can cause the choke to malfunction—or not work at all. When you put the carburetor on the manifold, operate the throttle before you tighten the flange nuts. Make sure the throttle opens and closes smoothly. Be positive the throttle linkage does not hang up on any part of the gasket or manifold.

Tubing vs. Hose Sizes—Tubing is typically measured by its outside diameter; hose is measured by its inside diameter. An easy way to remember this is that hose will fit over tubing of the same listed size: 3/8-inch hose slips over 3/8-inch tubing. Pipe is also measured by inside diameter.

Effect of air leakage on idle speed—Any air leak into the intake system affects idle speed and quality. Air leaks can be caused by worn valve guides, worn throttle-shaft bores in the carburetor, holes in any hose connected to intake-manifold vacuum, non-sealing gaskets where the manifold joins the cylinder head or carburetor. Leaks in any component attached to the manifold—such as vacuum motors, power-brake accumulators, etc., can also cause problems.

Fuel supply system—Read the fuel-system chapter for important details on fuel-lines, fuel filters and fuel pumps.

SORTING OUT THE GASKETS—Throttle bores of non-spread-bore Holley four-barrels range from 1-3/16 to 1-3/4 inch. Spread-bore 4165/4175's have 1-3/8 inch primary and 2 inch secondary bores. The 4360 has 1-3/8 inch primary and 1-7/16 inch secondary bores. Always check the gasket against the carburetor base. Make sure the gasket does not hang into the throttle bores to act as a restrictor or obstruct throttle operation. Some GM manifolds have heat tracks—others do not. The following info from the Chevrolet Parts Book and Holley's Performance Parts Catalog may help.

Square-flange Holleys on manifolds with heat tracks (thru 1969), use 0.010-inch stainless-steel heat shield GM 3884575 with GM 3890495 or 8BP-1331 gasket.

Square-flange Holleys on manifolds without heat track, use GM 3849825 or 8R-1077 thin gasket (0.020-inch-thick) for vacuum-secondary carburetors. Thin gasket can't be compressed much, so there is little chance of warping the base when you tighten the flange nuts. nuts.

1970 and later small-blocks with Holley (non-spread-bore), use 1/4-inch-thick hard-composition insulator/gasket GM 3999198 or 5/16-inch-thick 8BP-1392.

Square flange Holleys on manifolds without heat track, use GM 3881847 or 8R-1077 thin gasket (0.020-inch-thick) for vacuum-secondary types. 8BP-1446 can be used with mechanical-secondary carburetors if care is used in tightening so you don't warp the base.

Aluminum heat shield for Holleys (any non-spread-bore except 4500's) 7 x 13 inches, 1/4-inch thick at gasket area. Added 1/4-inch spacer (GM 3999198, 8BP-1392) required (or cutout) if high-capacity accelerator pump/s used.

Holley 4165/4175 or 4360 replacing Q-jet thru 1969 models, use 0.010-inch-thick stainless-steel heat shield GM 3884576 or 8BP-1567 against carburetor base. Use GM 3884574 or 8BP-1473 between heat shield and manifold flange.

Holley 4165/4175 or 4360 replacing Q-jet on 1970 and later models, use GM 3998912 or 8BP-1683—a gasket with phenolic bushings around stud openings.

Aluminum heat shield for 4165/4175 or 4360 with integral 1/4-inch-thick gasket area is GM 3969837. Requires cutout or 1/4-inch spacer (GM 3998912) to clear high-capacity accelerator pump at rear of 4165.

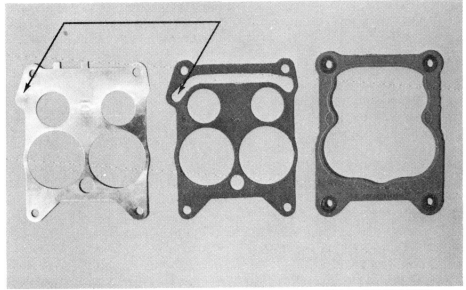

Heat shield surface marked TOP goes against carburetor base. This is GM 3884576 or Holley 8BP-1567. Center gasket, GM 3884574 or Holley 8BP-1473 fits between heat shield and manifold flange and must be placed so arrowed sections mate or there will be an exhaust leak. This arrangement was used on Chevrolets through 1969. 1970 and later eliminated the heat shield. Gasket/spacer at right, GM 3998912 or 8BP-1683 has phenolic bushings around stud holes. Note the heat shield in the photo below. One of these for spread-bore-type carburetors is GM's 3969837.

Two non-spread-bore Holley spacer/gaskets supplied as Chevrolet parts. At left is 3999198, a 1/4-inch-thick composition gasket used on Holley-equipped 1970-71 small-block Chevys. Aluminum heat shield 3969835 for 1970 big-blocks with Holleys has attached gaskets to give 1/4-inch thickness. Cutouts clear 3310 and Chevrolet and Holley double-pumpers. A 1/4-inch spacer is required between carburetor base and shield if a high-capacity accelerator pump is used. You could use 8BP-1392, a 5/16-inch-thick heat insulator. When using any gasket/spacer under the carburetor, be sure the openings do not restrict primary or secondary throttle bores or throttle operation.

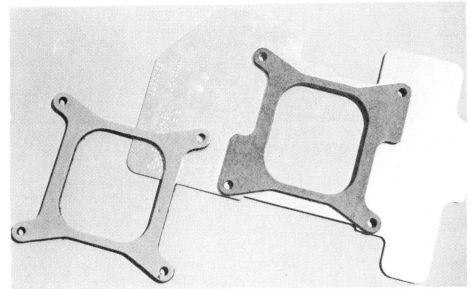

GASKETS FOR MODELS 2300, 4150/4160, 4165/4175 AND 4500 CARBURETORS

Several metering-block, metering-plate and fuel-bowl gaskets are used for these carburetors. Unfortunately, few are interchangeable. When disassembling your carburetor, keep the old gaskets soaking in a can of gasoline or solvent to compare to the new ones. Old gaskets (or new ones left in the sun) can get all out of shape if they dry out. Then it can be hard to figure out which ones you need. You want to be sure to replace the gaskets with the correct new ones. Holley Performance Parts Catalog lists correct gasket numbers for most popular carburetors. Holley Illustrated Parts & Specs Manual lists gasket and PEP repair kit numbers for all Holley carburetors. Some PEP kits contain more than one type of gasket so one kit can be used to service several carburetor numbers.

Some original-equipment gaskets are adhesive-treated for improved sealing. Extra effort may be required to disassemble carburetors with these gaskets.

To make things less confusing, the available gaskets are shown here with information on their general use. When buying gaskets in bubble packs, the number will include BP instead of R. For instance, five 8R-1023 gaskets are sold as bubble pack 8BP-1023. Where the part ends with RC, this is a resin-coated gasket recommended for street use anywhere high underhood temperatures are encountered.

Mating holes for locator pins on the metering blocks ensure that gasket openings will line up with those in the metering block, metering plate or fuel bowl. If the gasket looks like it doesn't fit, flop it over and try it another way.

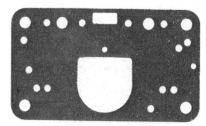

1 8R-1023 and 8R-1571-RC—Primary-metering-block gasket for most Model 4150, some 4160, early 4165, and most 2300 carburetors. It is also used as a secondary-metering-block gasket on double-pumpers. This gasket is not used where there is an accelerator-pump transfer tube. It is also used for Model 4500's without an intermediate system such as List 7320 and 4575.

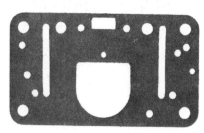

2 8R-1684-RC—Same as gasket 1, except with slotted idle-channel areas for improved operation where gasoline additives (low lead, detergents, etc.) and high heat cause distortion of the gasket in these areas. Improves idle operation in such cases.

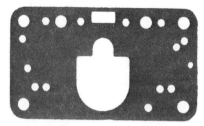

3 8R-1572-RC—Use on same carburetors as gasket 1 and 2 when they are equipped with an accelerator-pump transfer tube (shown on page 56). This gasket is used on all late-model 4165/4175 and a few 4150 carburetors. It is also used on the primary side of some 4160's. Gaskets 1 and 3 are not interchangeable, but a 3 can be made from a 1 or 2 by cutting a half circle in the center section to clear the transfer tube.

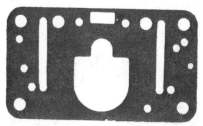

4 8R-1685-RC—Same as gasket 3, except with slotted idle-channel areas for improved operation where gasoline additives (low lead, detergents, etc.) and high heat cause distortion of the gasket in these areas. Improves idle operation in such cases.

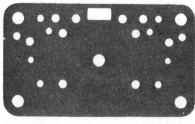

5 8R-1352—Secondary fuel-bowl gasket for Model 4160 and 4175 carburetors and diaphragm-operated outboard 2300's on 3 x 2 manifolds.

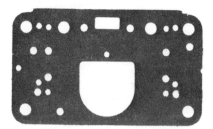

6 8R-1522—Primary and secondary metering-block gasket for Model 4500's with intermediate systems such as List 6214 and 6464.

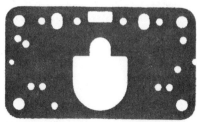

7 8R-1561—Metering-block gasket for Model 2300, List 6425 650 CFM two-barrel.

8 **8R-1594**—Primary metering-block gasket for Model 4160 Chrysler applications beginning in 1968.

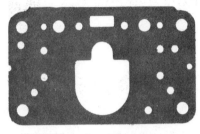

9 **8R-1576**—Primary metering-block gasket for Model 4160 Chrysler carburetors with accelerator-pump transfer tube. Gaskets 8 and 9 are not interchangeable, but you can make a 9 from an 8 by cutting out a half circle to clear the transfer tube.

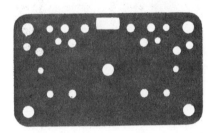

10 **8R-1174**—Secondary fuel-bowl gasket for Model 3160 three-barrel carburetor.

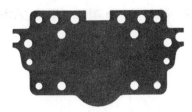

11 **8R-726**—Secondary metering-plate gasket for some Model 4160's.

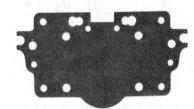

12 **8R-1175**—Secondary metering-plate gasket for Model 3160 three-barrel.

13 **8R-1437**—Secondary metering-plate gasket for Model 4160 Chrysler and outboard Model 2300 on some 3 x 2 applications with diaphragm-operated throttles.

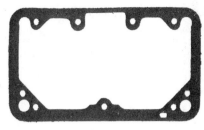

14 **8R-1022**—Fuel-bowl gasket for Models 2300, 4150, 4160, and 4500 carburetors.

15 **8R-1579**—Fuel-bowl gasket for all Model 4165 and some 4150, 4160 (primary side) and 2300. Also primary-bowl gasket for 4175. Differs from gasket 14 in that it works with the newer pump system with the discharge check in the metering block. Difference is the position of the slot (arrow).

NOTE—Both 14 and 15 have been changed so there is only one accelerator-pump hole. Earlier versions had two holes. Newer gaskets must be installed so the holes line up or there will be no pump shot.

Carburetor/Manifold Relationship

Holley Street Dominator manifold for Pontiac 326, 350, 400, 428 and 455 CID engines with or without EGR. Flange is drilled and tapped to accept either spread-bore or square-flange Holley carburetor.

BASIC DESCRIPTION

The intake manifold mounts the carburetor and provides connecting passages between the carburetor throats and the cylinder-head ports. A manifold has to divide the fuel, air, exhaust residuals and recirculated exhaust (EGR) equally among the cylinders at all speeds and loads. Each cylinder should receive the same amount of the fuel/air mixture with the same F/A ratio as all other cylinders to allow minimum emissions and for maximum power production.

As part of the design, passages should have approximately equal length, cross-sectional area and geometric arrangement. For various reasons, this cannot always be done, so flow-equalizing features are often used to make unequal-length passages perform as if they were nearly equal.

Because the manifold has to distribute both gas and liquid, the designer has to consider some liquid fuel is nearly always moving around on the walls and floor of the manifold. Controlling this by various devices such as sumps, ribs and dams is extremely important in ensuring equal F/A ratios for all cylinders. As an aid to vaporization and as part of controlling liquid fuel which has not been vaporized (or which drops out of the mixture due to an increase in absolute pressure), heating of the manifold is also essential, except for all-out racing manifolds. To avoid gravity influences which could cause uneven distribution of the liquid fuel, the manifold floor and carburetor base are typically parallel to the ground—instead of parallel to the crankshaft centerline which may be angled for drive-train alignment.

Small carburetor venturis aid fuel/air velocity through the carburetor and into the manifold and generally aid vaporization and hence distribution. High-speed air flow through the carburetor and manifold tends to maintain turbulence, but turbulence reduces flow. Velocity helps to maintain fuel droplets in suspension, thereby tending to equalize cylinder-to-cylinder F/A ratios. If the mixture is allowed to slow, as occurs wherever a passage size increases, fuel may separate from the air stream to deposit on the manifold surfaces. This causes variations in the F/A supplied to the cylinder.

Passage design should give sufficiently fast mixture flow to maintain good low and mid-range throttle response without reducing volumetric efficiency at high RPM. And, the internal design of the passages must carry the fuel/air mixture without forcing the fuel to separate from the air on the way to the cylinder head. When a mixture of fuel and air is forced to turn or bend, the air can turn more quickly than the fuel and the two may separate. When fuel separates from the air, a cylinder being fed with the mixture after the bend will receive a leaner F/A ratio than if the fuel had stayed with the mixture stream.

Pulsing within the manifold must also be accommodated. This phenomenon, also called *back-flow* or *reversion* by some, occurs twice during a four-cycle sequence. A pulse of residual exhaust gas enters the intake manifold during the valve-overlap period when both the intake and exhaust valves are off their seats. Another pulse toward the direction of the carburetor occurs when the intake valve closes. Either of these pulses may hinder flow in the manifold. In the most severe cases, the pulsing travels through the carburetor toward the atmosphere, and appears as a fuel cloud standing above the carburetor inlet. This is called "standoff."

Fuel is metered into the air stream *regardless of the direction of the stream* —down through the carburetor—or up through it from the manifold. The phenomenon occurs in some degree in most manifolds and in a marked degree in others—as we will discuss later in this section.

F/A ratio differences between cylinders are often remedied by using unequal jet sizes. Richer (larger) jets are used in a section of the carburetor feeding lean cylinders and leaner (smaller) jets are used for those carburetor barrels feeding rich cylinders. Although unequal jetting —commonly called *cross-* or *stagger-jetting* is a partially effective remedy—in such situations *the manifold is at fault.*

Stagger jetting is not very effective with plenum or single-plane manifolds because all cylinders share the output of all carburetor barrels.

Manifolds also provide mounts for other accessory devices, sometimes serve as tappet-chamber covers and may provide coolant passages or attachments. Provisions for choke operation, including a mount for a bi-metal thermostat and exhaust heat passages to this unit may also be included.

BASIC MANIFOLD TYPES

There are a number of different manifold types: Single-plane, two-plane, high-rise single- or two-plane, individual- or isolated-runner (IR) and IR with plenum chamber (Tunnel, Plenum or Channel Ram).

Manifold development is tough work. Here Holley engineer Ted Kartage (right) and technician Frank Walter were photographed as they engaged in dynamometer testing of one of the new Dominator manifolds on a 1975 455 CID Oldsmobile. Twenty test engines were used in the initial development of the Street, Street/Strip, and Strip only Dominator manifolds. Two dynos were used, both equipped with infra-red exhaust-gas analyzing units plumbed to each cylinder to measure cylinder-to-cylinder distribution accurately.

Types of V-8 manifolds

CROSS-H

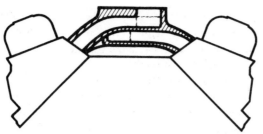

Cross-H or two-level manifold feeds half of cylinders from one side of carburetor — other half from other side of carburetor. Two sides of manifold are not connected.

SINGLE-PLANE

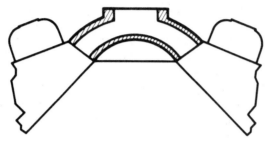

Single-plane manifold has all cylinder intake ports connected to a common chamber fed by the carburetor.

PLENUM-RAM

Plenum-ram manifold has a plenum chamber between passages to intake ports and carburetor/s.

ISOLATED-RUNNER

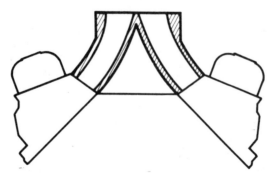

Isolated-runner (IR) manifold uses an individual throttle bore of a carburetor for each cylinder. There is no interconnection between the intake ports or throttle bores.

Single-Plane Manifolds—Roger Huntington, in a CAR LIFE article, "Intake Manifolding," said, "The very simplest possible intake-manifold layout would be a single chamber that feeds to the valve ports on one side and draws from one or more carburetor venturis on the other side. This is called a 'common chamber' or 'runner' *(single-plane)* manifold. Common-chamber manifolds have been designed for all types of engines—in-line sixes and eights, fours, V8's. It's the easiest and cheapest way to do the job. In fact, most current in-line four and six-cylinder engines use this type of manifold."

In the same article, Huntington also explained, "But we run into problems as we increase the number of cylinders. With eight cylinders there is a suction stroke starting every 90° of crankshaft rotation. They overlap. This means that one cylinder will tend to rob air-fuel mixture from the one immediately following it in the firing order, if they are located close to each other on the block. With a conventional V8 engine there are alternate-firing cylinders that are actually adjacent—either 5-7 on the left in AMC, Chrysler, GM firing order or 5-6 or 7-8 in the two Ford firing orders."

Two-plane (Cross-H) manifolds have been shown to be more throttle- and torque-responsive at low- and mid-range than most—but not all single-plane designs. As of 1970, performance manifolds of the single-plane variety began to appear in a new configuration which kept mixture stream speed reasonably high at low- and mid-range speeds, yet offered low restriction to flow at high engine speeds.

Holley's Dominator intake manifolds are exceptionally good examples of this type of single-plane design. Holley engineers found that careful selection of runner configuration, length and arrangement—coupled with optimum plenum volume—practically eliminated the low-speed disadvantage typically associated with single-plane manifolds. And, they were able to maintain the high-speed advantage over dual-plane-type manifolds.

Dual-runner (dual-port) manifolds appear to be two-plane designs, but they are two single-plane manifolds. A small- and a large-passage manifold are stacked one above the other in a single casting. The network of small passages connected to the primary side of the carburetor provides high-speed mixture flow for good distribution and throttle response at low- and mid-range RPM. Larger passages, connected to the secondary portion of the carburetor, supply extra capacity for high-RPM operation.

Experimental dual-runner *two-plane* manifolds have been developed by the automobile manufacturers. Development work on this concept was done by Ford, AMC, International Harvester, Holley Carburetor and the Ethyl Corp. Most of this work was oriented toward emission reduction. Volvo produced a manifold in the mid-sixties that used a similar idea. Mazda rotary (Wankel) engines use the same general manifolding idea.

Carburetor selection is thoroughly covered in that chapter, but observe that smaller carburetors can be used effectively with single-plane manifolds because the common chamber damps out most pulsing (which reduces flow capability). A small carburetor quickens mixture flow and thereby improves throttle response at low- and mid-range.

Two-Plane Manifolds—These are two single-plane manifolds arranged so each is fed from one half (one side) of a two- or four-barrel carburetor—and each of the halves feeds one half of the engine. A cross-H manifold for a V8 is a typical example familiar to most readers. Each half of the carburetor is isolated from the other—and from the other half of the manifold—by a plenum divider. Manifold passages are arranged so successive cylinders in the firing order draw first from one plane—then the other. In a 1-8-4-3-6-5-7-2 firing order, cylinders 1, 4, 6, 7 draw from one manifold plane and one half of the carburetor. 8, 3, 5, 2 are supplied from the other plane and the other side of the carburetor.

Because there is less air mass to activate each time there is an inlet pulse, throttle response is quicker and mid-range torque is improved—as compared to a conventional single-plane manifold.

The division of the manifold into two sections causes flow restriction at high RPM because only one half of the carburetor flow capacity and manifold volume are available for any intake stroke. Thus, the divider is sometimes reduced in height—or removed completely—to make more carburetor and manifold capacity available. This reduces bottom-end performance while top-RPM capability is increased because the volume increase on each side or section reduces mixture speed at low RPM. When the divider is removed, there is also some of the "robbing" effect mentioned earlier in the single-plane manifold discussion. This is not as severe as with a single-plane manifold, but it is a problem to be considered when modifying a two-plane manifold this way.

On engines built for high-RPM operation, divider removal can sometimes be a tuning plus. For street/track applications where the camshaft and other engine pieces are chosen to build torque into an engine, the divider should be left in place. The super-tuner can remove small amounts of the divider to reach the optimum divider height for a particular engine-parts combination and application.

Independent-runner Manifolds—Independent- or isolated-runner (IR) manifolds are race-only devices with one carburetor throat and one manifold runner per cylinder-head inlet port. Each runner/carburetor-throat arrangement is totally isolated from its neighboring cylinders. Carburetors used in IR setups have complete fuel-metering functions for each barrel because cylinders cannot share functions (except for accelerator pumps). There is a separate idle and main metering system for each barrel. An example is Holley's Model 4500, List 6214. An important benefit of an IR-type manifold is it allows tuning to take advantage of ram effect. The shorter the tuned length, the higher the torque peak.

Carburetors used in IR systems must be *much larger* than for other manifold types because each cylinder is fed by just one carburetor throat or barrel. Another reason for needing a larger carburetor is the severe pulsing in such systems reduces carburetor flow capacity in the RPM range where standoff becomes severe. In any IR system, some method of standoff containment should be used. The usual method is a stack long enough to contain the standoff atop the carburetor inlet.

Plenum-ram Manifold—This "almost-IR" manifold uses a plenum chamber between the carburetor base and manifold runners. This chamber helps to dissipate the strong pulsing so less of it enters the carburetor to disrupt flow. Just as important, it allows the cylinders to share carburetor flow capacity. In the typical dual-quad plenum manifold, three or four cylinders are simultaneously drawing mixture from a plenum being fed by eight carburetor bores.

The sharing of carburetor capacity allows the plenum-ram approach to manifolding to be "very forgiving" in terms of carburetor capacity so long as the carburetors are not too large for the RPM range to be used. In the case of the typical Pro-Stock drag racer, 6,800 to 8,500 RPM is usual. With a large-displacement engine, there is usually enough air flow at 6,000 RPM to start the main systems, even with a pair of the largest 4500's.

High-rise or high-riser? Optional high-rise high performance manifolds have been offered by some of the automobile manufacturers and by several aftermarket manifold makers, too. *High-rise* should not be confused with *high-riser*, which merely refers to spacing up the carburetor base so there is a longer riser available to straighten flow before it enters the manifold. The main function of a high-riser is to improve distribution by eliminating directional effects caused by a partially opened throttle, as described in the engine requirements chapter. A high-rise manifold, on the other hand, aligns the cylinder-head port angle with that of the manifold passage or runner. There is a nearly straight downward path from the carburetor to the intake port. In the case of a V8 engine, this usually means the entire network of manifold runners is raised—with the carburetor mounting.

Both high-rise and high-riser designs may be used with either single-plane or two-plane manifolds.

Classic cross-H manifold design was chosen by Ford engineers for their 427 CID single-overhead-cam drag-racing engine. Vacuum-secondary Holley four barrels were mounted "backwards." Ford Motor Company photos.

MULTIPLE CARBURETOR MANIFOLDS

Until about 1967, two or more carburetors were considered essential for a modified or high-performance engine. No self-respecting automotive enthusiast would consider building an engine with less than two carburetors unless the rules required it. But that situation changed rapidly over the past few years and now a single four-barrel carburetor is usual and accepted for *all* street—and many competition—applications. The availability of sophisticated two- and four-barrel carburetors with small primaries and large, progressively operated secondaries allow single-carburetor manifolds to provide excellent performance and low emissions, too.

The only application remaining for multiple carburetion—at least as we once knew it—is on drag cars and boats. And, it also fits into the extended high-RPM operation encountered at the Bonneville National Speed Trials. Multiple carburetion is used to reduce the pressure drop across the carburetors to an absolute minimum so that there will be the least possible HP loss from this restriction. Of course, low-RPM operation is not a consideration for such uses, so the extra venturi area or carburetor flow capacity does not create any problems.

A great many engines were sold with two- and three-carburetor manifolds as original stock equipment, but by 1973 all U.S. manufacturers converted their high-performance engines to single-four-barrel configurations. The vast majority of these special factory manifolds used Holley two- or four-barrel carburetors with vacuum-operated secondary throttles—or secondary carburetors.

Should you be the proud owner of a car with 2 x 4's or 3 x 2's, by all means avoid buying new mechanical linkages to open the throttles simultaneously. A lot of words scattered throughout this book explain the need for adequate mixture velocity and we have told you how to obtain it through selecting the correct carburetor capacity for the engine size and RPM. Simultaneous opening of multiple throttles (except for individual runner manifolds) goes completely against all of these recommendations, making the car hard to drive because velocity through the carburetors at low and medium speeds is drastically reduced —as compared to the more desirable progressively-operated secondaries.

HOLLEY DOMINATOR MANIFOLDS

In 1976, Holley introduced a totally new and complete line of intake manifolds. These manifolds resulted from one of the most intensive engineering-development programs ever undertaken by Holley. Dynamometer, emissions, fuel economy, street driving and drag-strip testing were all used to create these manifolds.

The development used the latest fuel-systems-engineering techniques. Runners were designed for minimum flow loss using cross-section-profile analysis. Infrared exhaust-gas-analyzing instrumentation accurately measured cylinder-to-cylinder fuel distribution, which was one of the paramount considerations in making manifolds which would not alter original equipment emissions performance.

Because Holley is the only manifold manufacturer which also makes carburetors and fuel pumps, they have been able to take a complete *induction-systems* approach. By considering the carburetor and manifold as a combination, Holley has been able to make several carburetor modifications to optimize induction-system performance.

By starting with a "clean slate," Holley engineers were able to create a line of manifolds with unusual features and capabilities. One feature found on the street manifolds is universal carburetor flanges which will mount either square-pattern or spread-bore carburetors without an adapter. Choke features are included so all carburetor operating features can be retained. All of Holley's street manifolds have full emission-control provisions. including EGR plumbing, plus adequate taps for operation of power brakes and other accessories. Exhaust-heat passages are provided to ensure fast warm-ups and good mixture distribution. Flange heights were kept as low as possible to fit under low hood lines.

During the development of the street manifolds, heavy emphasis was placed on making the manifold/carburetor combinations provide good power with stock cast-iron exhaust systems. Testing with headers ensured the manifold/carburetor teams would also work well with this type of equipment.

The "start-from-scratch" approach allowed Holley to incorporate a number of features allowing a minimum number

Big-block Chevy Strip Dominator. Typical of all Holley strip manifolds, there is no EGR provision, choke or exhaust heat. Note how the runners are isolated for low inlet-charge temperature and hence higher volumetric efficiency.

Big-block Chevrolet street manifold installed on a 1973 Corvette with an 800 CFM 4165. Car belongs to Holley engineer Ted Kartage, who was quite eager to try this early production sample.

Small-block Chevy Strip Dominator manifold with a 750 CFM double pumper on an A/SM car.

Dyno-power curve compares a Chevrolet LT-1 manifold and a competitive manifold. Carburetor was an R-4781 (850 CFM double-pumper). Each manifold is shown with its respective "best-power" jetting.

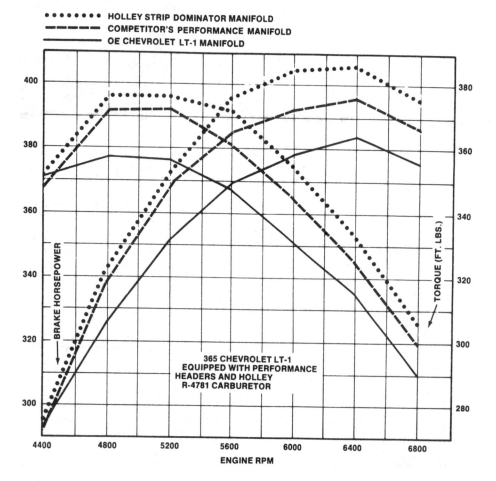

of manifolds to fit the maximum number of automobiles, trucks and recreational vehicles. Incidentally, all Holley manifolds are 100% pressure-tested. As much as possible, these manifolds allow the carburetor to be the torque-limiting device. This eliminates the need for several manifolds for specific RPM ranges with the same engine.

Considerable research went into the installation kits to make them the most complete ever offered. Installation directions are written so even inexperienced mechanics can install a manifold and make it work right.

Strip-type Dominator intake manifolds are designed for maximum power on fully modified competition engines. They are not designed for the street. There is no provision for chokes, EGR or other emission-control equipment. An air-gap between the manifold valley cover and the intake runners reduces heating the intake charge from engine heat. The result is increased density for the intake charge supplied to the cylinders and hence better power. These manifolds are true *racing* manifolds. Runner configurations and plenum volumes are optimized to favor all-out performance. Testing was done on the dynamometer and at the drag strip with competition engines.

The strip manifolds are designed for square-pattern Holley performance carburetors, except manifolds for Chrysler 318/340/360 CID engines work with both spread- and square-bore carburetors.

Development of the Strip Dominator was conducted with headers only and the engineers were primarily interested in power production above 4,800 RPM.

Performance comparisons between the Strip Dominator with an 850 CFM Holley R-4781 double-pumper carburetor and another make competition manifold showed a 12 HP advantage with the Holley manifold on a 355 CID small-block Chevrolet engine.

The manifolds introduced during 1976 were Phase I of an ongoing concentrated testing and development program. Additional manifolds of other types and to fit additional engines are planned for future release.

Street Dominator for Chrysler 318/340/360 CID engines. Flange accepts either spread- or square-bore carburetors.

Street Dominator for Ford 289/302 CID engines. EGR pad at rear can be plugged if not required.

Look real hard and you can see the Holley Street Dominator manifold in this 1975 Mustang II. The carburetor is a 600 CFM Model 4160. This package is a real mover. Photo shows complexity of late-model vehicles. The combination shown here passed all emission requirements.

A Holley Street Dominator manifold on a 318 CID Chrysler gave over 30 HP gain in both cases, but headers kept the power up at higher RPM.

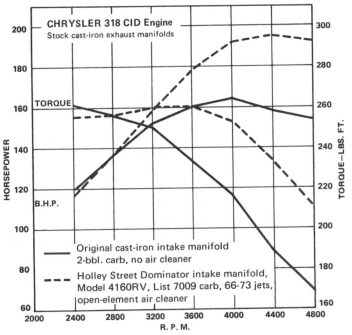

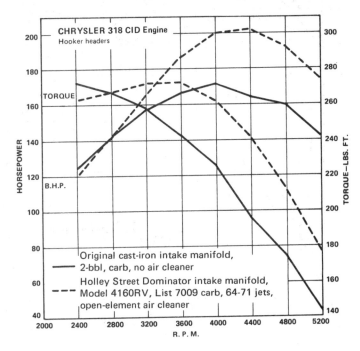

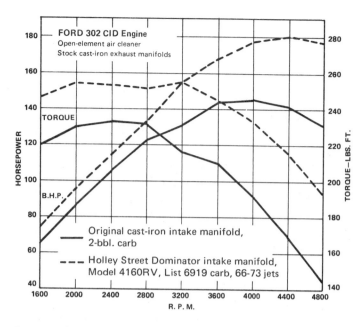
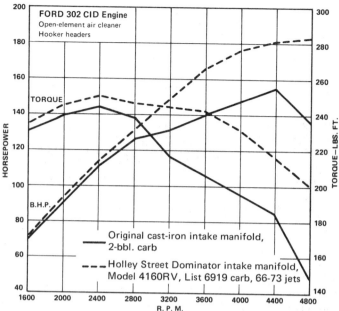

Between the stock intake and exhaust manifolds and a Holley Street Dominator manifold there was about 35 HP difference vs. around 27 HP difference with headers. Headers gave slightly more power and carried power further up the range.

Looking down the throat of a deep-breathing Street Dominator. Note the straight path the inlet charge has from the carburetor to the cylinder.

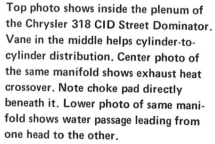

Top photo shows inside the plenum of the Chrysler 318 CID Street Dominator. Vane in the middle helps cylinder-to-cylinder distribution. Center photo of the same manifold shows exhaust heat crossover. Note choke pad directly beneath it. Lower photo of same manifold shows water passage leading from one head to the other.

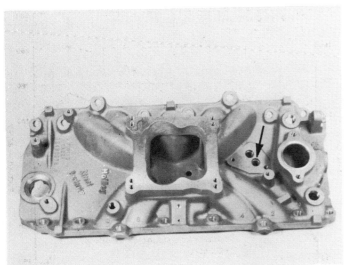

Big-block Chevrolet Street Dominator. Universal mounting pad accepts either square or spread-bore carburetors without an adapter. EGR provision (arrow) is in all Holley street manifolds. Overall height was kept as low as possible for installation under low hoods. Photo at right shows clamp-on type EGR valve on big-block Chevy street manifold. Universal pad also accepts bolt-on valves.

EGR pad on the small-block Chevy street manifold. Plugs are provided in the installation kit to seal the passages on applications where no valve is required. Plugs drive in easily and a little sealant helps prevent leaks.

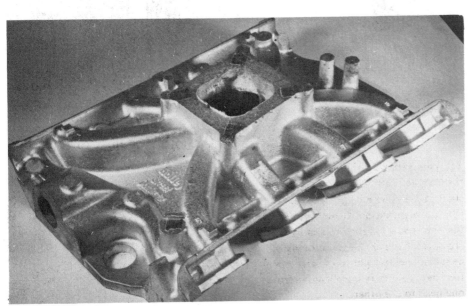

Aluminum casting prior to machining. This is Holley's Street Dominator manifold for the Ford 352, 390, 428 engines with or without EGR. You'll need a crane to get the original equipment manifold off the engine as it weighs about 100 pounds.

MANIFOLD/CARBURETOR APPLICATIONS

Strip Manifolds	Carburetor
Chevrolet Small-Block	780 CFM with diaphragm secondaries or 800 CFM double-pumper if class allows.
Chevrolet Big-Block (Oval or Rectangular ports)	850 CFM double-pumper or Model 4500 with adapter.
Chrysler Small-Block (318–360 CID)	780 CFM with diaphragm secondaries or 800 CFM double-pumper if class allows.
Ford 351 C	780 CFM with diaphragm secondaries or 800 CFM double-pumper if class allows.

Street Manifolds	
Chevrolet Small-Block	Model 4360 for maximum economy. 600 CFM with diaphragm secondaries as best compromise for all-round performance. 650–750 CFM double-pumper for best performance. Either spread-bore or square-flange carburetors can be used without adapter.
Chevrolet Big-Block	600 CFM with diaphragm secondaries for good economy and mid-range performance. 650–800 CFM double-pumpers for best performance. Either spread-bore or square-flange carburetors can be used without adapter.
Ford Small-Block (289–302 CID)	Model 4360 for maximum economy. 600 CFM with diaphragm secondaries for economy and over-all performance. 600–700 CFM double-pumpers for best performance. Kickdown linkage must be considered for automatic transmission versions.
Ford 390–428 CID	Use 600–780 CFM with diaphragm secondaries for best all-round performance. 750–800 CFM double-pumpers for maximum performance. Manifold will not accept spread-bore without an adapter. Kickdown lever must be considered for automatic transmission versions.
Ford 351 C	Use 600 CFM with diaphragm secondaries for best all-round performance. 750 CFM double-pumper for maximum performance. Accepts spread-bore carburetors without an adapter. Kickdown lever must be considered for automatic transmission versions.
Chrysler Small-Block (318–360 CID)	Model 4360 for maximum economy. 600 CFM with diaphragm secondaries for economy and good all-round performance. 600–700 CFM double-pumpers for best performance. Manifold accepts either spread-bore or square-flange carburetors.
Chrysler 383–440 CID (two manifolds)	600 CFM with diaphragm secondaries for best economy and all-round performance. 780 CFM with diaphragm secondaries or 750 CFM double-pumper for best performance. Manifold accepts spread-bore carburetors without an adapter.
Pontiac 326–455 CID	Model 4360 for maximum economy. Best all-round performance with diaphragm-secondary. 600–780 CFM, depending on engine displacement. 650–800 CFM double-pumpers for best performance. Manifold accepts either spread-bore or square-flange carburetors.
Oldsmobile Small-Block	Model 4360 for best economy. 600 CFM with diaphragm secondaries for best all-round performance. 650–750 CFM double-pumpers for maximum performance. Manifold accepts either spread-bore or square-flange carburetors.
Oldsmobile Big-Block	Model 4175 or 780 CFM with diaphragm secondaries for best all-round performance. 750–800 CFM double-pumper for best performance. Manifold accepts either spread-bore or square-flange carburetors.

This chart gives manifold applications available as of mid-1976 with recommended carburetor or carburetors by size and configuration. Part numbers were left out to keep the chart from becoming obsolete. See the latest Holley Performance Catalog for the right part numbers for manifolds and carburetors. For street applications you'll want to choose the numbers that are emission-tested and labeled as "emission/performance" in the catalog.

Chrysler low-block 361, 383, 400 CID engines can get a performance boost with this Holley Street Dominator manifold designed for use with either spread-bore or square-flange Holley carburetors.

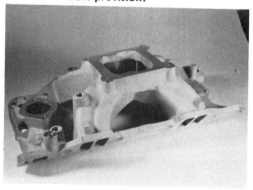

Holley Strip Dominator manifold for Chrysler small-block V-8 engines: 318, 340 and 360 CID. Runners on this manifold are kept separate from the cam/tappet cover so they will stay as cool as possible for a dense charge. There is no EGR provision.

Holley's manifold installation kits are the most complete you'll ever see. You may find parts not needed for your particular engine. The "extra" parts are there to ensure a simple trouble-free installation on any application specified in the Performance Parts Catalog.

Carburetor & Engine Variables

So many things affect the operation of the carburetor and the engine that it seemed important to put these *variables* into one chapter where you could read about them. Hopefully, our words have placed these in the proper relationship so you can begin to see how each variable affects others. Much of this information is truly important in understanding and tuning stock and racing engines. Study it and you'll be the local expert.

SPARK TIMING

Late-model cars have labels in the engine compartment which specify engine idle speed and spark-advance settings. These have been carefully worked out by the factory engineers to ensure that the engine in that vehicle will have emissions within specified limits. The trend is to use a retarded spark at idle and during the use of the intermediate gears. Some smog systems which lock out vacuum advance in the intermediate gears have a temperature-sensitive valve which brings straight spark advance into operation if the engine starts to overheat and during cold operation prior to the time that the engine water temperature reaches 80°F.

Most carburetors provide ports for timed spark and for straight manifold vacuum. Some Ford applications use a combination of venturi vacuum and manifold vacuum, called a "pressure spark system." There was a trend away from this combination in 1972.

In general, a retarded spark reduces oxides of nitrogen by keeping peak pressures and temperatures at lower values than would be obtained with an advanced spark setting. It also reduces hydrocarbon emissions. This is good for reducing emissions—but bad for economy, drivability and heating of the coolant. Fuel is being burned in the engine, but some energy is wasted as heat flows into the cylinder walls and as excess heat in the exhaust manifolds—which also aggravates the problem of heating in the engine compartment. The cooling system has to work harder. In essence, fuel is still burning as it passes the exhaust valve. Thermal efficiency of the engine is less because there is more wasted energy.

The use of a retarded spark requires richer jetting in the idle and main systems to get off-idle performance and drivability. A tight rope is being walked here because the mixture must not be allowed to go lean or higher combustion temperatures and a greater quantity of oxides of nitrogen will be produced. Also, if mixtures are richened too far in the search for drivability, CO emissions increase.

Because retarding the spark hurts efficiency, the throttle plate must be opened further at idle to get enough mixture in for the engine to continue running. This fact must be considered by the carburetor designer in positioning the idle-transfer slot. It must also be considered because high temperatures at idle and high idle-speed settings definitely promote dieseling or after-run.

Dieseling is caused primarily by the greater throttle opening, but it is aggravated by the higher average temperatures in the combustion chambers. Higher temperatures tend to cause any deposits to glow so that self-ignition occurs. On a car equipped with an anti-dieseling solenoid, turning the ignition on causes the solenoid to move the throttle to its idle setting. When the ignition is turned off, the solenoid retracts the idle setting about 50 RPM slower than a normal idle setting (a more closed idle position for the throttle) so that the engine will not have a tendency to run on. If it can't ingest enough mixture to continue running, the dieseling is usually stopped effectively.

There is also a Combined Emission Control (CEC) solenoid on 1971 and later GM cars. This is not an anti-dieseling solenoid. It operates on a different principle for accomplishing a different approach to the emission-reduction problem as described in the emissions chapter.

On carburetors equipped with timed spark advance (no advance at closed throttle) a port in the throttle bore is exposed to vacuum as the throttle plate moves past the port—usually slightly off-idle. The spark-advance vacuum versus air-flow calibration is closely held in the production of the carburetor because this has a dramatic effect on HC and NO_X emissions. The distributor advance, once held to be so important for economy, has now become an essential link in the emission-reduction chain.

VALVE TIMING

Valve timing is the one item with the greatest effect on an engine's idling and low-speed performance. Adding valve overlap and lift with a racing cam allows the engine to breathe better at high RPM, but it worsens manifold vacuum at idle and low speeds—creating distribution and vaporization problems. These are obvious because the engine becomes hard to start, idles very roughly—or not at all—has a bad flat spot coming off idle and has very poor pulling power (torque) at low RPM. This is especially true when a racing cam is teamed with a lean, emission-type carburetor.

Because manifold vacuum is reduced, the signal available to pull mixture through the idle system is also reduced, causing a leaning of the mixture. Also, the throttle will have to be opened further than usual

---

Dieseling—This is the tendency of an engine to continue running in an irregular and rough fashion after the ignition has been switched off. This problem is aggravated (1) by anything which remains hot enough to ignite fuel/air mixtures—such as any sharp edges in the combustion-chamber area and (2) by the high idle-speed settings used to meet emission requirements.

---

to get enough mixture into the engine for idling. This can place the off-idle slot/port in the wrong relationship to the throttle so that there is insufficient off-idle fuel to carry the engine until main-system flow begins.

And, the idle mixture has to be made richer to offset the poor vaporization and distribution problems. Part of the poor distribution and poor drivability problem stems from the overlap period. When both the exhaust and intake valves are open simultaneously, some of the exhaust gas which is still in the cylinder at higher-than-atmospheric pressure pushes into the intake manifold to dilute the incoming charge. This charge dilution effectively lowers the combustion pressure. This is especially true up to that RPM where the overlap time interval becomes short enough so that the reverse pulsing becomes insignificant.

Reverse pulsing through the venturis at wide-open throttle causes added fuel flow, creating a richer mixture. Once the main system starts, the discharge nozzle will deliver fuel in response to air flow in either direction. More information on valve timing, reverse pulsing and so forth is contained in the ram-tuning section of the following chapter. Read it carefully!

When the manifold vacuum is reduced (pressure increased toward atmospheric), the power valve may start to operate, or at least to open and close as the manifold vacuum varies wildly. This is not a valid reason to remove the power valve, but it will require selecting a power valve which will be closed at the lowest vacuum (highest pressure) which occurs during idling.

In extreme cases, a "wild" racing cam may magnify these problems to the point where the car becomes undrivable for anything except a competition event. This is especially true if the carburetion capacity has been increased to match the deep-breathing characteristics of the cam.

There have been many cases where a racing cam was installed at the same time that the carburetor was changed. If the mechanic did not understand what was occurring in the engine—and many do not understand—the carburetor was often blamed for poor idling and the bad flat spot off idle when the real culprit was the racing camshaft! Detailed information on tuning an engine with a racing camshaft is contained in the performance-tuning section of the next chapter. Refer to page 131.

TEMPERATURE

Temperature greatly affects carburetion. If affects mixture ratio because air becomes less dense as temperature increases (1% for every 11°F). The density change reduces volumetric efficiency and power, even though main jet corrections of approximately one main jet size smaller for every 40°F ambient temperature increase can be made to keep mixture ratio correct.

Maximum power production requires that the inlet charge be as cold as possible. For this reason, racing engines are usually designed or assembled so that there is no exhaust gas heating of the intake manifold. Stock-type passenger-car and truck engines, on the other hand, use an exhaust-heated intake manifold because the warmer mixture—although not ideal for maximum power—helps drivability. A warm-air inlet to the carburetor and/or an exhaust hot spot in the intake manifold greatly aid vaporization. Good vaporization ensures that the mixture will be more evenly distributed to the cylinders because fuel vapors move much more easily than liquid fuel.

Icing occurs most frequently at 40°F and relative humidity of 90% or higher. It is usually an idle problem with ice forming between the throttle plate and bore. It usually occurs when the car has been run a short distance and stopped with the engine idling. Ice builds up around the throttle plate and shuts off the mixture flow into the engine so that the engine stops. Once the engine stalls, no vaporization is occurring, so the ice promptly melts and the engine can be restarted. This may occur several times until the engine is warm enough to keep the carburetor body heated to the point where vaporization does not cause the icing. We have also seen cases of "turnpike icing" in the venturi itself when running at a relatively constant speed for a long period of time under the previously stated ideal icing conditions. In this case, ice build-up chokes down the venturi size so that the engine runs slower and slower. Vaporization of fuel removes vast quantities of heat from the surrounding parts of the carburetor, hence there is a greater tendency for this phenomenon to occur where vaporization is best, namely, in small venturis. Icing is no longer a major problem because most cars are factory-equipped with exhaust-manifold stoves to warm the air which is initially supplied to the air cleaner inlet. Thermostatic flapper valves (on some engines) shut off the hot-air flow when underhood temperatures reach a certain level.

On some high-performance engines, a vacuum diaphragm opens the air cleaner to a hood scoop or other cold-air source at low vacuums/heavy loads.

At the other end of the thermometer we find a phenomenon called percolation. It usually occurs when the engine is stopped during hot weather or after the engine has been run long enough to be fully "warmed up." Engineers call this a "hot soak." In this situation, no cooling air is being blown over the engine by the fan or vehicle motion and heat stored in the engine block and exhaust manifolds is radiated and conducted directly into the carburetor, fuel lines and fuel pump.

Ice can cause damage—Because the accelerator pump is the lowest point in the fuel-inlet system of a Holley with removable fuel bowls—water can collect there —particularly in a carburetor which is not used regularly. If this water freezes, using the carburetor could rupture the diaphragm. To avoid such problems by purging water from the entire system, add a can of a "dry gas" compound to absorb water in the system so that it will be carried into the engine with the fuel. This should be done at the start of the cold-weather season.

Fuel in the main system between the fuel bowl and main discharge nozzle can boil or "percolate" and the vapor bubbles push or lift liquid fuel out of the main system into the venturi. The action is quite similar to that which you have seen in a percolator-type coffee pot. The fuel falls onto the throttle plate and trickles into the manifold. Some excess vapors from the fuel bowl—and from the bubbles escaping from the main well—are heavier than air and also drift down into the manifold. This makes the engine very difficult to start and a long cranking period is usually required. In severe cases, enough fuel collects in the manifold so that it runs into cylinders with open intake valves. The fuel washes the oil off of the cylinder walls and rings and severe engine damage can result.

Percolation is aggravated by fuel boiling in the fuel pump and the fuel line to the carburetor because this can create fuel-supply pressure as high as 15 to 18 psi—sufficient to force the inlet valve needle off its seat. Fuel vapor and liquid fuel are forced into the bowl so that the level is raised. This makes it that much easier for the vapor bubbles to lift fuel to the spill-over point.

Solving the percolation problem requires several solutions. The main system is designed so that vapor bubbles lifting fuel toward the discharge nozzle tend to break before they can push fuel out of the nozzle. Fuel levels are carefully established to provide as much lift as can be tolerated.

In some instances, the fuel will be made to pass through an enlarged section at the top of the main well or standpipe to discourage vapor-induced spillover.

Gaskets and insulating spacers are used between the manifold and carburetor and between the carburetor base (throttle body) and fuel bowl. In some instances, an aluminum heat deflector or shield is used to keep some of the engine heat away from the carburetor.

As mentioned in the distribution section of the engine requirements chapter, an abnormally warm day in winter when fuel typically has a high vapor pressure will cause hot-starting problems due to percolation.

Another way that hot-starting problems are reduced is through the use of internal bleeds in the fuel pump and/or a vapor-return line on the carburetor ahead of the inlet valve. When the bleed and/or return line are used, any pressure buildup in the fuel line escapes harmlessly into the fuel tank or fuel-supply line.

Another bad effect caused by high temperature is boiling of the fuel in the fuel line between the pump and the fuel tank. Fuel can even boil in the fuel pump itself. When the fuel pump and line are filled with hot fuel, the pump can only supply a mixture of vapor and liquid fuel to the carburetor. Because very little liquid fuel is being delivered during an acceleration after a hot soak, the fuel level drops, causing leaning. In fact, the bowl may be nearly emptied, partially exposing the jets. When the jets are partially exposed, the carburetor cannot meter a correct fuel/air mixture because the jets are designed to work with liquid fuel—not a combination of liquid and vapor.

This condition is called vapor lock. Holley carburetors are specially immune to the problem because their large bowl capacity helps to provide an adequate supply of liquid fuel until the pump can supply liquid fuel, even though a portion of the fuel has escaped as vapor.

In extreme cases, fuel lines may have to be rerouted to keep them away from areas of extreme heat, such as the exhaust system. If the lines cannot be relocated, it is usually possible to insulate them. This is especially important for high-performance on racing vehicles. Cool cans help, as do high-performance electric fuel pumps located at the tank to *push* fuel to the carburetor.

Formula racing cars are often seen with fuel radiators or coolers.

AIR DENSITY

In the air requirements section we related volumetric efficiency of the engine to the density of the fuel/air mixture which its cylinders receive. We showed that the higher the density, the higher the V.E.

The density of the mixture depends on atmospheric pressure which varies with altitude, temperature and weather conditions. And, the mixture density is also affected by the intake-system layout. Density is increased when the carburetor is supplied with cool air and when the intake manifold is not heated. Density is reduced if the inlet air is heated or if the fuel/air mixture delivered by the carbu-retor is heated as it enters the manifold. Further reductions in density occur as the mixture picks up heat from the manifold and cylinder-head passages, hot valves, cylinder walls and piston heads.

As with almost all other engine variables, there are tradeoffs. Warming the mixture reduces its density—but also improves distribution, especially at part throttle.

Density has an effect on carburetor capacity and mixture. Because the *major* density changes which occur are due to altitude changes, let's consider the effects of driving from dockside at Los Angeles to the 5000-foot altitude of Denver. The flow and pressure-difference expressions look like this:

$$Q \sim A \sqrt{\frac{\Delta p}{\gamma}} \quad \text{or} \quad \Delta p \sim \left(\frac{Q}{A}\right)^2 \times \gamma$$

where

Q = volume flow

γ = density

Δp = pressure difference

A = carburetor-venturi area

Regardless of density, the volume taken in remains the same at a given RPM, but going from sea level to 5000 feet drops density to 83% of its sea-level value. Pressure difference or Δp is also 83% of the original value so the carburetor acts as if it were larger. Because pressure change or Δp increases inversely as the square of the area, area must be reduced only by a ratio of $\sqrt{.83}$ or .91 to restore the same Δp as experienced with the original carburetor at sea level. In other words, the carburetor acts as if it were 9% larger. This is only a problem if the carburetor size was marginally too big to begin with.

If the carburetor was on the verge of stumbling and flat spots in acceleration at sea level—due to late turn-on of the main system, those problems will be worse at high altitudes because of the *effective* increase in carb size due to reduced air density.

When the carburetor is flowing enough air at high altitudes to turn on the main system, the fuel/air ratio will be rich, for a related reason.

We used the following equation in Chapter 2 to show that WOT fuel/air ratio is determined mainly by venturi size and main jet size—assuming fuel and air densities never change.

$$\frac{M_f}{M_a} = \frac{A_{mj}}{A_v} \times \frac{K_f}{K_a} \times \frac{\sqrt{\gamma_f}}{\sqrt{\gamma_a}} \times \frac{\sqrt{\Delta P_{mj}}}{\sqrt{\Delta P_v}}$$

where

M_f = mass flow of fuel
M_a = mass flow of air
A_{mj} = area of main jet
A_v = area of venturi
K_f = constant
K_a = constant
γ_f = density of fuel
γ_a = density of air
ΔP_{mj} = change in pressure across main jet (same as ΔP_v)
ΔP_v = change in pressure across venturi (same as ΔP_{mj})

We use the equation again to show what happens when air density *does not* stay the same. The change is only in the term:

$$\frac{\sqrt{\gamma_f}}{\sqrt{\gamma_a}}$$

If γ_a changes to 0.83 of its original value, $\sqrt{\gamma_a}$ changes to 0.91 of its original value, a decrease of 9%. The end result is an *increase* in fuel/air ratio of 9%, meaning the mixture contains 9% more fuel for the same weight of air. The engine runs rich; fuel economy and power are reduced.

A rule of thumb is to reduce the Holley jet size by one number for each 2000-foot altitude increase. Holley engineers usually figure that there is approximately 4% fuel-flow change between jet sizes. Thus, two jet sizes smaller would be required to keep nearly equivalent fuel/air ratios at the 5000-foot altitude.

COMPRESSION RATIO

High compression improves engine performance by increasing the burning rate of the fuel/air mixture so that peak pressures and peak torque can approach the maximum of which the engine is capable. High compression also increases emissions of hydrocarbons and oxides of nitrogen.

Lowering the compression reduces HC emissions by reducing the surface-to-volume ratio of the combustion chamber. The greater the surface-to-volume ratio,

the more surface cooling which occurs, thereby increasing HC concentration.

Until about 1970, high-compression engines with up to 11:1 c.r. were available in high-performance cars. By 1971, manufacturers were reducing compression ratios and by 1972, most cars had no more than 8.5 or 8:1 c.r. By reducing compression, the burning rate of the fuel/air mixture is slowed so that peak pressures which enhance the formation of NO_x cannot be reached. Reduced compression also increases the amount of heat which is transferred into the cylinder walls because more burning is still going on as the piston is descending, thereby raising the exhaust temperature.

The use of low compression increases the fuel/air requirement at idle because the greater amount of residual exhaust gas remains in the clearance volume and combustion chamber when the intake valve opens, causing excessive dilution of the mixture. This can cause off-idle drivability problems. In effect, low compression provides a degree of exhaust-gas recirculation (EGR) without plumbing or hardware as described in the emissions chapter.

Raising or lowering the compression ratio of an engine does not normally affect the main-system fuel requirements, so jet changes are not usually required when such modifications are made. Raising compression may require slightly less ignition advance in some cases.

Compression ratio and octane requirement are closely related. As compression is increased, octane must also be increased (a higher-octane fuel used) to avoid detonation and preignition—often referred to as "knock." Similarly, lowering an engine's compression reduces its octane requirement.

After World War II, engines were designed and produced with high-compression designs to obtain higher efficiencies. By 1969, some had been available with 11 to 1 compression! Then the need for reduced emissions began to be approached on the basis of getting ready for the future.

First, the auto makers asked the fuel companies to start "getting the lead out" in preparation for the coming emission equipment which would not be able to tolerate lead in the exhaust.

Engineers have created engines which

will operate on lead-free gasoline. Tetraethyl lead (Ethyl compound) is one of the most commonly used anti-knock additives used as an octane-increaser in gasolines. Lead compounds are being studied for their effects on plant life, the atmosphere and humans—and there is a serious move to rid these from exhaust on an ecological basis. But, this is not the real reason that engineers want to eliminate lead from fuel, as previously mentioned. The expensive catalyst used in catalytic mufflers—standard parts in 1975-76 cars—is literally destroyed when contaminated with lead and lead by-products. Thus, the automotive engineers had to ensure that there will be little likelihood of the users being able to buy gasoline which could wreck the catalytic mufflers when they are used. These and other emission devices are discussed in the emission chapter.

In 1971, engines with lower compression ratios became standard items in all lines. And, by 1972, most engines were being made with only 8 to 1 c.r. The gasoline refineries have been steadily reducing gasoline octane and the trend is to still lower octanes. Where 100+ octane gasolines were commonplace in the late sixties, 90-octane was standard by 1975.

EXHAUST BACK PRESSURE

Exhaust back pressure has very little effect on carburetion at low speeds. For high performance, low back pressure is desired to obtain the best volumetric efficiency through optimum breathing. The effects of exhaust restriction increase approximately as the square of the RPM.

Using headers may change the main-jet requirement. Either richer or leaner jets may be needed, depending on the interrelation of the engine components.

Carburetor & Performance

When we made this chapter different from the preceding one on engine variables it was difficult to decide which chapter should include what subjects. Because the variables covered in the previous chapter do affect performance, be sure to read both of these chapters if you are working to obtain maximum performance.

RAM TUNING

Ram tuning can give better cylinder filling—and thus improved volumetric efficiency—in a narrow speed range by taking advantage of a combination of engine-design features:

Intake and/or exhaust-system passage or pipe lengths

Valve timing

Velocity of intake and exhaust gases

Although ram tuning improves torque at one point or narrow RPM band, the improvement tends to be "peaky" so that the power falls off sharply on either side of the peak. It is generally understood that ram tuning is a resonance phenomenon. Resonance is sought at a tuned peak with the knowledge that *the power gained at that point may be offset by a corresponding loss at other speeds.*

Ram tuning can be used to add mid-range torque—as was done on some of the Chrysler six-cylinder and V-8 engines in the early 1960's. These gains are obtained at the expense of top-end power. Or—more usual for high-performance engines—low- and mid-range torque may be sacrificed to take advantage of top-end improvements.

Individual intake-manifold passages for each cylinder (ram or tuned length) can be measured from the intake valve to a carburetor inlet if there is one venturi per cylinder. Or, the length may be measured from the intake valve to the entry of a plenum chamber fed by one or more carburetors. This is a function of manifold design. *The longer the passage length, the lower the RPM at which peak torque will occur.* Although equations have been written to describe where ram-tuning effects will

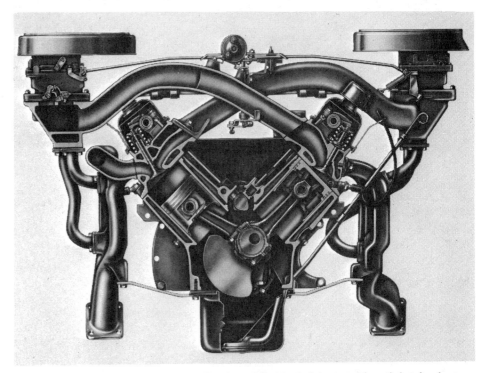

1963 Chrysler 413 CID 300J Ram Induction engine had eight equal-length intake ducts. Each four-cylinder set was fed by one carburetor and an equalizing tube connected the two sets. Duct lengths were selected to give best torque improvement at 2800 RPM— about 10% better than conventional manifolding. The strong torque increase provided a noticeable acceleration improvement over a 1500-RPM range. . .from 50 to 80 MPH. This engine is considered to be a classic example of ram-tuning use in a production car.

occur, most of these have been over-simplifications which leave out the effects of manifold-passage size and the sizes of intake ports and valves. Making any of these larger raises the RPM at which best filling occurs. This explains why best drivability and street performance is obtained with small-port manifolds.

Valve timing greatly affects the RPM capabilites of an engine. Cylinder filling can be aided at high RPM by holding the intake valve open past BC (Bottom Center). Consider what happens. At low speeds, holding the valve open past BC allows part of the intake charge to be blown back into the intake manifold as the piston rises on its compression stroke. This reduces manifold vacuum and drastically affects idle and off-idle F/A mixture requirements, as described in the engine-

variables chapter.

As RPM is increased, faster piston movement creates a greater pressure drop across the carburetor. Air enters the carburetor with higher velocity, giving greater acceleration—and momentum—to the mixture traveling toward the valve. Thus, as the piston approaches BC on the intake stroke, cylinder pressure is rising toward that at the intake port. And, pressure at the intake port is being increased by air-column momentum in the intake-manifold passage supplying it. Thus, filling improves with RPM until friction losses in the manifold exceed the gain obtained from delayed valve closing.

The past two paragraphs are true, regardless of whether ram tuning is being used or not. Now let's consider what is happening in the manifold passage as the

Chrysler's production plenum-ram for the hemi engine is constructed to provide peak power around 6800 RPM. Branch length and plenum volume are varied to fit specific racing applications. This manifold is the hemi version of a similar manifold supplied for large-block wedge engines. Two Holley List 3116 vacuum-secondary carburetors with 1-11/16-inch throttle bores are used.

More about ram tuning — Complete details on ram-tuned intake systems are in Philip H. Smith's book, *The Scientific Design of Exhaust & Intake Systems.* Perhaps one of the best single articles written on the subject was by Roger Huntington in the July 1960 Hotrod magazine, "That Crazy Manifold." July and August 1964 Hotrod magazines had two further articles by Dr. Gordon H. Blair, PhD. All are worth reading. However, the use of a dynamometer to measure the real performance of a specially constructed "tuned" system — either intake or exhaust — is absolutely essential.

Blocked heat risers and cold manifolds are race-only items — Blocked heat risers may prove acceptable for street driving in the summer, but when it starts to get chilly, change back to the exhaust-heated manifold. Otherwise, the long time required for the engine to warm the manifold for acceptable vaporization and smooth running will make the car miserable to drive. Getting the heat working again plays an important part in making the engine last longer. If you insist on feeding a poorly vaporized mixture to the cylinders during cold weather, excess gasoline will wash the oil from the cylinders and cause very fast engine wear. For street use the small loss in volumetric efficiency is not usually important because the engine is usually much larger than it needs to be anyway!

valve is opened and closed. When the piston starts down on its intake stroke, a rarefaction or negative-pressure pulse is reflected to the carburetor inlet. As this pulse leaves the carburetor, atmospheric pressure rushes in as a positive-pressure pulse. When the passage length is optimum for the RPM at which peak torque is sought, the positive-pressure pulse arrives near the time when the valve is closing, assisting in the last part of cylinder filling. Note the interrelationships of passage length, mixture velocity, valve timing and RPM. It's a complex process, to say the least!

The mixture attains a velocity of up to 300 feet per second or more (depending on RPM) as it travels through the port during the intake stroke. Because the mixture has mass (weight), it also has momentum which is useful for aiding cylinder filling when the intake valve is held open after BC (sometimes to 100° past BC) while the piston is rising on the compression stroke. Manifold design enters in at this point because an isolated-runner (IR) system allows much longer delay in closing the intake valve than the usual single- or two-plane manifolds. This is because the IR system supplies only one cylinder per carburetor venturi. There is no other cylinder to consider from the standpoint of mixture dilution, sharing (robbing) or adverse pulsing caused by another cylinder.

When the intake valve shuts, incoming mixture piles up or stagnates at the valve backside, reflecting a compression (positive-pressure) pulse or wave toward the carburetor inlet. As this wave leaves the carburetor inlet, it is followed by a negative-pressure pulse back to the valve. This bouncing or reflective phenomenon repeats several times until the inlet valve again opens. Pressure at the carburetor inlet varies from positive to negative as the wave bounces back and forth in the inlet passage.

At certain engine speeds, the reflection or resonance phenomenon will tend to be in phase (in synchronism) with the intake valve opening and closing so that the positive pulses will tend to "ram" the mixture into the cylinder for improved filling.

Although the pulsations are in phase only at certain speeds (yes, there can be multiple peaks!), the mixture column in the intake passage provides some ram effect at all speeds due to its own inertia.

Best filling is obtained with the intake-system and exhaust-system resonances in phase at the same RPM.

On the exhaust side, exhaust gas enters the pipe at a pressure of 80 psi or higher because the exhaust valve is opened before BC—before the conclusion of the power stroke—while there is still pressure in the cylinder. Thus, the exhaust gets a "head start" so the piston does not have to work so hard in pushing out the exhaust. The exhaust "pulse" starts a pressure wave traveling at the speed of sound (in hot gas) to the end of the system. From the end of the system, a rarefaction or low/negative-pressure pulse is reflected back to the exhaust valve at the same speed. Tuned systems are constructed with lengths to allow this pulse to arrive during the overlap period to ensure complete emptying of the cylinder. The idea here is to reduce exhaust residuals in the clearance volume. This reduces charge dilution and provides more volume for F/A mixture—hence greater volumetric efficiency.

Main jet requirements with ram tuning

With an IR manifold, ram tuning has a dramatic effect on main-jet requirements because the strong pulsing at wide-open throttle pulls fuel out of the discharge nozzle in *both* directions. This is not of any consequence if part-throttle performance is of no concern—because main-jet size can be established for WOT operation. Otherwise, the F/A mixture must be richened for part-throttle operation. An inverted power valve can be used to accomplish this. Or, a plenum chamber added to the manifold or carburetor base will soften the high-RPM WOT pulsing which allows a compromise jetting to be used. This will usually handle both part-throttle and WOT operation.

While not a strictly IR application, the Volkswagen with a center-mounted two-barrel has an extreme pulsing condition because only two cylinders draw from each venturi. In the Buggy Spray carburetor, an inverted power valve is used to offset the lack of pulse-caused enrichment at part-throttle. And, a plenum slot in the base helps to "soften" WOT pulsing.

Components of a fresh-air hood are clearly shown in this photo of a Camaro. Chevrolet engineer Gerry Thompson holds the air-cleaner base assembly which fits onto the two four-barrel Holleys. Hood is open at back by cowl. Duct in hood mates with foam rubber gasket. Smaller photo shows how hood opening overhangs cowl.

Increased density means larger jets—No matter how a density increase is obtained —whether by increased atmospheric pressure, a cold manifold or a cooler inlet-air temperature—the increase must be accompanied by the use of larger main jets. The size increase (in area) is directly proportional to the square root of the density increase (in percent).

You can calculate relative air density from barometer and temperature readings, but most tuners prefer an air-density meter. This can be used to select the correct main-jet area according to the square root of air-density changes (in percent). Air density changes from hour-to-hour and day-to-day—and most certainly from one week to the next and one altitude to another. These are available from K & D Accessories, Box 267-H, Longview, WA 98632 and from Moon Equipment Co., Santa Fe Springs, CA.

Dual snorkles or inlets are usually one of the trademarks of a high-performance vehicle as it comes from the factory. One above is a 1972 Pontiac GTO. Lower photo shows Chevrolet Z-28 air-cleaner arrangement. This high-performance air cleaner is one of the least restrictive single-element cleaners in the business.

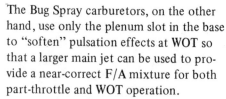

The Bug Spray carburetors, on the other hand, use only the plenum slot in the base to "soften" pulsation effects at WOT so that a larger main jet can be used to provide a near-correct F/A mixture for both part-throttle and WOT operation.

The bigger the carburetor in relation to the engine size, the greater the need for an inverse power valve. The smaller the carburetor in relation to the engine size, the less difference there will be between part-throttle and WOT F/A mixture.

Tuned-exhaust effects on the main-jet requirement vary. If the exhaust causes stronger intake-system pulsing, a smaller main jet could probably be used. Or, if the effect is to lessen pulsing, a larger jet might be required. Predicting these effects is extremely difficult, so tuners tackle each situation on an individual cut-and-try basis.

Summary of ram tuning

The peaky results obtained with ram tuning can seriously reduce engine flexibility and this point must not be overlooked. It is all too easy to get over-excited about the spectacular results obtained from racing motorcycle engines, so let's look at that area briefly before leaving the subject of ram tuning.

Motorcycle engines are typically built as single-cylinder units, each with its own carburetor and exhaust pipe. This allows the designer to take advantage of two very important features: (1) No mixture-distribution problems, and (2) pressure phenomena in the intake and exhaust systems can be relied on for ramming or "supercharging" the cylinder at a desired RPM—usually quite high.

There's nothing very mysterious about a fuel-cooling can (cool can). Fuel passing through the coiled line is cooled by ice or dry ice and alcohol in the can. Fuel temperature is lowered to ensure that carburetor/s receive liquid fuel. Cold fuel under pressure is not likely to flash into vapor when it enters the carburetor and sees only atmospheric pressure.

Thus, very-high-output bike engines obtain HP outputs of up to 2.9 HP per cubic inch! However, racing motorcycles with engines of this ilk have 8- to 11-speed gearboxes to allow using the very narrow RPM band in which power is produced. Some of these engines will not produce noticeable power below 6000 RPM!

RAM AIR

The forward motion of your car can induce some air pressure into the intake system—provided there is a forward-facing inlet connected to the carburetor. This duct should preferably have air straighteners in it so air enters the carburetor smoothly—preferably through an air cleaner or other diffusing device which will break up any turbulence.

The scoop should mate with a tray under the carburetor/s so any pressure achieved is not wasted into the entire engine compartment.

Very minor pressure increases are obtained in this way, but even a minor pressure can aid the induction system at high RPM and give more HP where the engine is starting to "run out of breath." According to Gary Knutsen of McLaren Engines, pressures of 6 inches of water (about 0.22 psi) were obtained at racing speeds on the Chapparal race cars. On very long straights this pressure provided measurable performance benefits as opposed to cars running without the pressurized air inlet.

Other writers have claimed the improvement can amount to as much as +1.2% at 100 MPH, +2.7% at 150 MPH and +4.8% at 200 MPH. No matter what the capability of an engine is, such increases could make the difference between winning or losing.

You may have noted Formula 1 and Formula 5000 cars nearly all run some form of forward-facing air scoop over the induction system. These have the scoop in the airstream above the driver's head. Drag-race cars with air scoops typically have the scoop opening ahead of the hood or else far enough above the hood so the scoop is not picking up turbulent air.

COLD AIR & DENSITY

Density of the mixture has been thoroughly discussed in the engine-requirements chapter where we showed that higher density inlet air improves the engine's volumetric efficiency proportionately to the density increase. Improving density with the use of cool inlet air and a cool inlet system was also mentioned in the previous chapter on engine variables. So, let's examine the practical aspects. What can you do to keep the density "up" to get the best HP from your engine?

First, the underhood temperature is not ideal for HP production because—even on a reasonably cool day—air reaching the carburetor inlet has been warmed by passing through the radiator and over the hot engine components. Underhood temperatures soar to 175°F or higher when the engine is turned off and the car is standing in the sun. An engine ingesting warm air will be down on power by more than you might imagine. Assume that the outside (ambient) air temperature is 70°F and the underhood temperature is 150°F. Use the following equation:

$$\gamma_{oa} = \frac{460 + t_{uh}}{460 + t_{oa}} \times \gamma_{uha}$$

$$\gamma_{oa} = \frac{460 + 150}{460 + 70} \times \gamma_{uha}$$

$$\gamma_{oa} = 1.15 \, (\gamma_{uha})$$

where

γ_{oa} = outside air density

γ_{uha} = underhood air density

t_{oa} = outside air temperature

t_{uh} = underhood air temperature

In the example here, outside-air density is 115% of the underhood-air density—or 15% greater. Because mass air flow increases as the square root of density, HP with outside air will increase $\sqrt{1.15}$ or 1.072 or 7.2%. If the engine produces 300 HP with 150°F air-inlet temperature, it can be expected to produce 322 HP with 70°F air-inlet temperature. So, the density increase afforded by using outside air is considerable.

Cold air gives more improvement than ram air because approximately 1% HP increase is gained for every 11°F drop in temperature. This assumes the mixture is adjusted to compensate for the density change and that there is no detonation or other problems. Using outside air instead of underhood air is climate-limited because too-cold temperatures may lead to carburetor icing.

If air scoops are being used to duct air to the carburetor, do not connect the hose or scoop directly to the carburetor. Instead, connect the scoop to a cold-air box or to the air-cleaner housing to avoid creating turbulence of the incoming air as it enters the carburetor air horn.

If you plan to use a cold-air kit which picks up cold air at the front of the car, be prepared to change the air-cleaner filter element at regular intervals—perhaps as often as once a week in dustier areas. If you leave the filter out of the system, plan on new rings or a rebore job in the near future because your engine will quickly wear out. You'll be better off ducting cold air from the cowl just ahead of the windshield. This is a high-pressure area which will ensure a supply of cool outside air to the carburetor. That area still gets airborne dust, but it is several feet off of the ground—away from some of the grimy grit encountered at road level. If you want to keep your car "looking stock," using fresh air from the cowl is another way to get performance without making the car look like a racer.

Cars with stock hoods and stock or near-stock-height manifolds may be equipped with fresh-air ducting to the air cleaner by using parts from GM cars such as some of the high-performance Chevrolets.

Or, you may prefer to use one of the fresh-air hoods which have been offered on some models. These typically mate a scoop structure on the hood with the air-

TABLE OF VELOCITY STACK EFFECTS
For Four-Barrel Carburetors

Carburetor	No Stack	3-1/2" Stack	5" Stack	Improvement
Model 4500, List 4575	1005 CFM			baseline
		1030		+2.4%
Model 4150, List 4781	825			baseline
		830		+0.6%
			833	+0.9%
Model 4165, List 6262	812			baseline
		845	845	+4%
				Holley Tests July, 1971

AIR CLEANER COMPARISON
Model 4165 — List 6210

Cleaner Type	WOT Air Flow (CFM)
None	713
Chevrolet 396 closed-element with single snorkle	480
Chevrolet 396 closed-element with single snorkle cut off at housing	515
Same as above, but with two elements	690
Chevrolet high-performance open-element unit	675
Same as above, but with two elements	713
14-inch diameter open-element accessory-type air cleaner	675
Chevrolet truck-type element (tall) used with accessory-type base and lid	713
Foam-type cleaner (domed flat-funnel type)	675

NOTE: All data obtained with same carburetor. New clean paper elements used in all cases.

Holley Tests
October 1971

cleaner tray on the carburetor/s. Thus, the underhood air is effectively prevented from entering the carburetor.

If a tall manifold such as an IR or plenum-ram type is being used, the hood will have to be cut for clearance and a scoop added to cover the carburetor/s. Whether the scoop is open at the front or back depends on the air flow over the car. The optimum entry for air into the scoop may have to be determined by testing. An optimum entry will provide an air supply which is above atmospheric pressure and is non-turbulent. In general, the area of the scoop opening should be approximately 12% larger than the area of the carburetor venturis. The roof of the scoop should be positioned 1-1/2 inch above the carburetor inlet. Any more clearance may create detrimental turbulence—any less will restrict air flow into the carburetor.

Like the fresh-air hoods, scoop-equipped hoods can be mated with a tray under the carburetor/s to ensure that warm underhood air cannot enter the carburetors to reduce inlet-air density.

Drag racers will always keep the hood open and avoid running the engine between events so the compartment stays as cool as possible. It is helpful to spray water onto the radiator to help cool it. This ensures that the engine water temperature will be lower and that the radiator will not heat incoming air any more than is absolutely necessary.

Although "seat-of-the-pants" feel may indicate stronger performance from a cold engine, the real fact of the matter is that the engine coolant temperature should be around 180°F or so to allow minimum friction inside the engine. The idea is to keep the engine oil and water temperature at operating levels while taking care to keep down the inlet-air temperature.

While we are talking about density improvement by using cold inlet air we should remember that a heated manifold reduces density. The exhaust heat to the manifold should be blocked off to create a "cold" manifold when performance is being sought after. There are various ways of accomplishing this. In some instances there will be intake-manifold gaskets available which close off the heat openings. Or, a piece of stainless steel or tin can metal can be cut to block off the opening when slipped between the gasket and manifold.

Many competition manifolds have no heat riser and therefore the manifolds are "cold" to start with. Some car makers offer shields which fit under the manifold to prevent heating the oil with the exhaust heat cross-over passage. These shields also prevent the hot oil from heating a cold manifold—or at least reduce that tendency. This can be a cheap way to gain approximately 10 HP on a small-block V-8 engine.

AIR CLEANERS

Because every engine needs an air cleaner to keep down expensive cylinder wear which is caused by dust, it makes sense to use one which will not restrict the carburetor's air-flow capabilities. By avoiding restrictions in the air cleaner, you allow your engine to develop full power.

The only time that an engine might possibly be run without an air cleaner would be on drag-car or drag-boat engines being operated where there is no dust in the air or pit—a very unlikely situation, to say the least. Even then, when an air cleaner is removed from the carburetor, the air-cleaner base should be retained because its shape ensures an efficient entry path for the incoming air and helps to keep the incoming air from being heated by the engine.

The accompanying table shows the results of tests which dispel some of the common fallacies about air cleaners and their capabilities. In general, a tall, open-element air cleaner provides the least restriction. It also increases air-inlet noise. Note that some of the air cleaners will allow full air-flow capability. These are the air cleaners which should be used by racers, even if their use requires adding a hood "bump." An air cleaner which gives full flow capability to the carburetor provides impressive top-end power improvements, as compared to one which restricts flow. For instance, the use of two high-performance Chevrolet air cleaners stacked together—instead of one open-element cleaner—improved a 1969 Trans Am Camaro's lap times at Donnybrook, Minnesota by one full second.

It is very important to check the clearance between the upper lid of the air cleaner and the top of the carburetor's pitot or vent tubes. Air cleaner elements vary as much as 1/8 inch in height due to production tolerances. The shorter elements can place the lid too close to the

pitot tubes so that correct bowl reference pressures are not developed. Whether the pitot tubes are angled or flat on top, there should always be at least 3/8-inch clearance between the tip of the pitot tube and the underside of the air-cleaner lid.

You have noticed the long "snorkle" intakes on modern air cleaners. These are put there to reduce intake noise—not to improve performance. High-HP engines nearly always have two snorkles for more air and perhaps for "more image," too. For competition, the snorkle can be removed where it joins the cleaner housing. Additional holes can be cut into the cleaner housing to approximate an open-element configuration to improve breathing. Or, it is sometimes possible to expose more of the element surface by inverting the cleaner top. As shown in the accompanying table, an open-element design is least restrictive.

When it comes time to race, use a clean filter element. Keep the air cleaner base on the carburetor if you possibly can, even if you have removed the air cleaner cover and element. But, be sure to secure the base so that it cannot vibrate off to strike the fan, radiator or distributor.

Because the carburetor is internally balanced, that is, the vents are located in the air horn area, no jet change is usually required when the air cleaner is removed.

VELOCITY STACKS

Velocity stacks are often seen on racing engines. These can improve cylinder filling (charging) to a certain extent, depending on a great many other factors. For instance, when velocity stacks are used on an isolated-runner manifold, the stacks may form part of a tuned length for the air column.

Velocity stacks also provide a straightening effect to the entering air. And, they can contain fuel standoff—which is typical with isolated-runner-design manifolds. Remember that the velocity stacks need space above them to allow air to enter smoothly. Mounting a hood or air-box structure too close to the top of the stacks will reduce air flow into the carburetor. Two inches should be considered a bare minimum clearance between the top of a velocity stack and any structure over it.

Note the table on page 121 for more details.

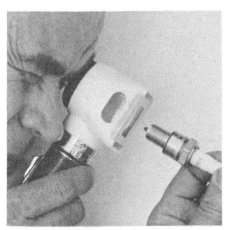

Color and appearance of plug porcelain can be checked with a spark-plug illuminator. These usually include a magnifier to aid in plug reading. One of these should be in every tuner's tool kit. The engine must be "cut clean" by turning off the ignition and declutching or getting into neutral at the conclusion of a wide-open-throttle full-power run in top gear. If this is not done, plug readings are apt to be meaningless.

PLUG COLOR CHART

Rich—Sooty or wet plug bases, dark exhaust valves.

Correct—Light-brown (tan) color on porcelains, exhaust valves red-brown clay color. Plug bases slightly sooty (leaves a slight soot mark when plug base turned against palm of hand). New plugs start to color at base of porcelain inside shell and this can only be seen with an illuminated magnifying viewer for plug checking. For drag engines with wedge (quench-type) combustion chambers, speed and elapsed time become more important than the plug color. Plugs may remain bone-white with best speeds/times. A mixture which gives the light-brown color may be too rich for these engines.

Lean—Plug base ash-grey. Glazed-brown appearance of porcelains may also indicate too-hot plug heat range. Exhaust valves whitish color.

NOTE: Piston-top color observed with an inspection light through the plug hole can be a quicker and sometimes more positive indicator of mixture than plug appearance. Careful tuners look at ALL indicators to take advantage of every possible clue to how the engine is working.

When an air cleaner is used on a carburetor equipped with stacks, keep the two-inch recommended clearance between the top of the stacks and the underside of the air-cleaner lid. This may require using two of the open-element air cleaners which have been "glued" together with Silastic RTV or other sealant—and a longer stud between the carburetor top and the cleaner lid.

Holley carburetors which have been designed for use with velocity stacks, such as the 600 CFM two-barrel, can sometimes obtain as much as seven to nine percent improvement in air flow by using stacks. On the usual four-barrel with the 5-inch air cleaner base, the use of wide-mouthed entry devices provided improvements which ranged from insignificant (less than 1%) to as much as 4%.

DISTRIBUTION CHECKING

Distribution checking to determine whether all cylinders are receiving an equal mixture becomes extremely important when a manifold or carburetor change is made. Even though a previous carburetor/manifold combination may have provided nearly perfect distribution you cannot take the chance that one or more cylinders will be running lean. Although several ways of checking distribution are detailed in the distribution section of the engine requirements chapter, the man at the race track usually has only one way to do it. He must rely on the appearance and color of the plug electrodes and porcelains. These can provide a lot of valuable information about what is happening in the engine.

An accompanying plug-color chart explains some of the things to look for. Because the part of most interest is the base of the porcelain—which is "buried" in the plug shell—a magnifier-type illuminated viewer should be an early purchase for your tool box.

Before continuing, we need to mention that checking plug color gives only a rough idea of what is occurring in the way of mixture ratio and distribution. Ford engineer Don Gonyou claims it is difficult to see a change in plug color without changing the main jet at least four sizes. And, from the list of items which cause plug-appearance variations (later in this section), it is easy to see that using plug color to check distribution will only be

helpful when the engine is in good condition. Engine condition can be checked quickly with a compression gage to make sure that all cylinders provide equal compression at cranking speed. Or, for a more accurate check, a leak-down test can be used to compare cylinder condition.

Remember that new plugs take time to "color"—as many as three or four drag strip runs may be needed to "color" new plugs. Plug color is only meaningful when the engine is declutched and "cut clean" at the end of a high-speed full-throttle, high-gear run. If you allow the engine to slow with the engine still running, plug appearance will be meaningless. Plug readings can be made after full-throttle runs on a chassis dyno with the transmission in an intermediate gear so that the

Chevy high-rise manifold for Z-28/LT-1 has oil shield under heat-riser area. By using this shield in conjunction with a blocked heat riser, the intake charge is cooled considerably!

Manifold Vacuum vs. RPM

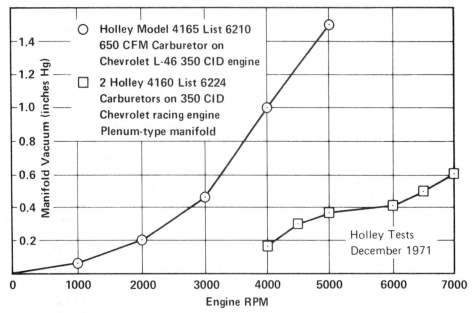

Chart above shows the minimum restriction provided by using two four-barrel carburetors for a drag-race or competition engine. This minimum flow restriction allows such installations to provide horsepower equivalent to that of racing fuel-injection systems in nearly all instances.

dyno is not overspeeded—but road tests require high gear to load the engine correctly.

Similarly, plug checks can be made where the engine has been running at full-throttle against full load applied by an engine dyno. It is much easier to get good plug readings on the dyno because full power can be applied and the engine cut clean without difficulty. Plugs can be read quickly because you can get to them easier than in the usual car or boat installation. However, don't think that plug heat range and carburetor jetting established on an engine dyno will be absolutely right for the same engine installed in your racing boat or chassis. Air-flow conditions past the carburetor can easily change the requirements—perhaps so unevenly that different cylinders will need different changes.

It would be nice if every plug removed from an engine looked like the others from the same engine—in color and condition—but this is seldom ever achieved! Color and other differences indicate combustion-chamber temperatures and/or fuel/air ratios are not the same in every cylinder—or that related engine components need attention. The problem is greatly complicated in engines where there is a

great difference between the cylinders in turbulence and efficiency. The big-block Chevrolet is a notable example as described in H. P. Books volume on that engine.

If differences exist in the firing end of the plugs when you examine them, the cause may be due to one or more factors: Unequal distribution of the fuel/air mixture, unequal valve timing (due to incorrect lash or a worn cam), poor oil control (rings or excess clearance). Problems within the ignition system which can also lead to plugs not reading the same or misfiring include: Loose point plate, arcing in the distributor cap, defective rotor, cap or plug wires/connectors, cross fire between plug wires, defective primary wire or even a resistor which opens intermittently.

Pay special attention to cleanliness of the entire ignition system including the inside and outside of the distributor cap and the outside of the coil tower. Also, clean the inside of the coil and cap cable receptacles. Any dirt or grease here can allow some of all of the spark energy to leak away.

If the cylinders have equal compression and the valves are lashed correctly, a difference in plug appearance may indicate that there is a mixture-distribution problem. It is sometimes possible to remedy

this with main-jet changes. For instance, if one or more plugs show a lean condition, install larger main jets in the throttle bore/s feeding those cylinders. Should one or more plugs appear to show a rich condition, install smaller main jets in the throttle bore/s feeding those cylinders. The real problem occurs when several cylinders fed from the *same* throttle bore show different mixture conditions—some lean with some correct—or some rich with some correct—or perhaps a combination of all three conditions! This shows a manifold fault which cannot be corrected with jet changes. Correcting such conditions requires manifold rework which is beyond the scope of this book.

HEADER EFFECTS vs. JETTING

Headers will usually reduce exhaust back pressure so that the engine's volumetric efficiency is increased (it breathes easier!). The main effects of headers will be seen at wide-open throttle and high RPM.

Using headers may change the main-jet requirement. Either richer or leaner jets may be needed, depending on the interrelation of the engine components.

The preceding ram-tuning section details how a tuned exhaust system which alters the pulsing seen by the carburetor may make main-jet changes necessary to compensate for any increase/reduction in intake-system pulsing.

CARBURETOR RESTRICTION

Carburetors are tested at a given pressure drop at Wide-Open Throttle (WOT) to obtain a CFM rating indicative of flow capacity. One- and two-barrel carburetors are tested at 3.0-inch Hg pressure drop. Three- and four-barrel carburetors are tested at 1.5-inch Hg pressure drop.

When you want to make comparisions, use these formulas:

$$\text{Equivalent flow at 1.5 in. Hg} = \frac{\text{CFM at 3.0 in. Hg}}{1.414}$$

$$\text{Equivalent flow at 3.0 in. Hg} = \text{CFM at 1.5 in. Hg} \times 1.414$$

The one- and two-barrel rating was adopted because low-performance engines typically showed WOT manifold-vacuum readings of 3.0-inch Hg. When four-barrel carburetors and high-performance engines became commonplace they were rated at 1.5-inch Hg pressure drop because of *two* reasons. First, this rating was close to the WOT manifold vacuum being seen in these engines. Secondly, although we've never seen this mentioned anywhere else, most of the carburetor-testing equipment had been designed for smaller carburetors. The pump capacity on this expensive test equipment was not adequate to provide 3.0-inch Hg pressure drop through larger carburetors. 1.5-inch Hg was about the limit of the pump capacity. Hence the 1.5-in. rating came about as a "happy accident."

For maximum output, it is essential to have the carburetor as large as possible—*consistent with the required operating (driving) range.* The driving range must be considered as discussed in the chapter on selection and installation.

Using a lot of flow capacity—more than calculations indicate necessary for the engine—can reduce inlet-system restriction and increase volumetric efficiency at WOT and very high RPM. Such carburetion arrangements can compete with fuel injection in terms of performance because the restrictions are minor. However, it should be noted that the dual-quad installations typically used by professional drag racers are not capable of providing usable low- or mid-range performance. These engines are typically operated in a very narrow range of 6000 to 8500 RPM or so.

TOOLS REQUIRED

Start with patience! You need more than the feel in the seat of your pants and the speedometer in your car to do serious tuning. Specific tools are required, but you'll also need a patient and methodical approach to the project. This further means you cannot be in a hurry. If you have no intentions of really getting serious about tuning, run your Holley as it comes out of the box and leave your tool box locked up.

A vacuum gage, fuel-pressure gage and a stop watch are essential. So is a tachometer. The vacuum and fuel-pressure gages are an extra set of "eyes" to let you look inside of the engine to see what is happening there. For serious competition on a regular basis, an air-density gage can be especially helpful. You'll also want a 1-inch open-end wrench, preferably the

MAC S-141 (specifically designed for the fuel-inlet nuts on two- and four-barrel carburetors with center-hung floats). A broad-blade screwdriver can be used for jet changes.

TIMING DEVICES

Although a stop watch can be used for some very fine tuning, you may want to consider getting your own timing device with associated photocells, or arrange for the rental of a portable unit so that you can set up the lights at varying distances. Pro Stock racers often test acceleration over an initial 60 feet to work out starting techniques, tire combinations and carburetion. Most races are won in the critical starting period and by the initial acceleration over the first few feet.

A FEW PARTS WILL BE HELPFUL

When you are working on a two- or four-barrel with detachable fuel bowls, you should have a few parts available. Foremost among these are extra bowl and metering-block gaskets and bowl-screw gaskets. O-Rings for the transfer tube should also be on your shopping list. Be especially careful to buy the correct bowl gaskets for carburetors with the non-pull-over pump discharge nozzles (such as the 4165's or the 6425). Although similar to those used on 4150/60 and 2300, they are NOT the same.

Before you buy any main jets, find out what size is already in the carburetor. If the carburetor is new, have the salesman check Holley's Variable Specifications and Parts catalog to see which jets are supposed to be in the carburetor. Or, check the Holley High-Performance Catalog. However, neither is foolproof because Holley changes jet sizes to fit specific application requirements and catalogs must be printed ahead of the time when some of the carburetors are released in final form for manufacturing. To be absolutely sure, take the top off of the carburetor or pull off the fuel bowls and look for yourself.

In the case of two- and four-barrel carburetors with removable fuel bowls, this means rechecking the accelerator-pump adjustment for each pump when you put the bowls back on. Make sure that the primary pump lever moves instantly as you operate the throttle lever and that there is at least 0.020 inch additional travel for the pump lever when the throttle

is wide open. The pump lever should contact the operating lever adjustment screw at idle.

Once you know what jets you have, buy four sizes lean and four sizes rich for each main jet in your carburetor.

This will handle most engine variations and most atmospheric-condition changes (density). Main jets are not necessarily the same size in a four-barrel. Secondaries may contain different jets than the primary side. And, if you should be running a big-block Chevrolet with an open-plenum service-package manifold, the carburetor could use three different jet sizes because it is "stagger-jetted." In the case of plenum-type all-out drag-race manifolds, install at least two sizes larger main jets. These are the only two cases where jets should be changed *before* running the engine. Otherwise, make your first tests with the carburetor jetted *exactly as supplied by Holley.* There are no other exceptions.

For any four barrel with a 650 through 850 CFM main body, the main jets are usually between 76 and 81, assuming that the power valves are installed in the carburetor if they were there originally. If the carburetor you are working with is a four barrel in this size range and the jets are more than four sizes away from the 76-81 spread—chances are that whoever jetted the carburetor lost his bearings and got off course with the jetting. Get it back to the original jetting before you start tuning.

Drag racing often requires richer jets. For instance, if 80's worked fine on the dynamometer, you may need 82's or 84's at the strip. Jets have to be rich to get a mixture equivalent to that obtained on the dynamometer with rock-steady conditions. Double-pumpers or carburetors with a high-capacity pump may not need extra-rich jetting for the strip because these pumps usually inject enough fuel to cover up the air-flow lag.

ONE THING AT A TIME

It never occurs to some would-be tuners—even some old-timers—that changes must be made ONE at a time. It's too easy to be tempted, especially when you are sure that you need a heavier flywheel, different gear ratio, other tires, a different main jet, two degrees more spark advance and a different plug heat-range. Changing one of these at a time would be just TOO SLOW.

You—or your buddies—convince yourself of that. And, first thing you know, you've lost the rabbit and you don't know where you're at or how you ever got there. When you are trying to tune your carburetor—a complex piece of equipment in itself—there's all the more reason to make one change at a time and then check it against known performance by the clocks, or by your stopwatch and tachometer as described elsewhere in this chapter.

Anytime that you change more than one thing, you no longer have any idea of which change helped—or hindered—the vehicle's performance. It's even possible that one change provided a positive improvement which was cancelled by the negative effects of the other change. The net result *seemed to be* no change.

When tuning, follow a procedure and stick to it. The less help (and therefore, advice) that you have, the better. Concentrate. Be deliberate. And don't be surprised when the process eats up more time than you ever thought it could have. If you are planning other changes soon, such as a different manifold, camshaft, air cleaner, distributor-advance curve, cylinder heads or exhaust system—put off tuning until the car and engine are set up as you expect to run it. Otherwise, your tuning efforts will be wasted and you can look forward to a repeat performance of the entire tuning process.

Whatever you do, don't fall into the common trap of rejetting the carburetor to some specialized calibration that you read about in a magazine article or heard about at the drag strip last week. Holley spent thousands of dollars getting the calibration correct and there is good reason to believe that they used more engineers, technicians, test vehicles, dynamometers, flow benches, emission instrumentation and other equipment and expertise than you may have available as you start your tuning efforts. Remembering these hard facts will save your time, effort, money and temper.

When you have made jet changes from standard, use a grease pencil or a marking pen to mark the bowls with the main-jet sizes which you've installed. Some tuners use pressure-sensitive labels on the fuel bowls for noting which jets are installed. Others write the information on a light-colored portion of the firewall or fender-

well where it won't be wiped off during normal tuning activities such as plug changes, etc.

Noting what jets have been used saves a lot of time when tuning because no time is wasted in disassembly and reassembly to see what jets are in the carburetor/s. Even the best memories are guaranteed to fail the jet-size memory test.

BEFORE YOU START TUNING

Now that you've gathered the tools, spare jets and other paraphernalia, there are a few more details to take care of. First and foremost, check the accelerator pumps to make sure they operate with the slightest movement. And, check that there is at least 0.015 to 0.020 added travel left in the diaphragm-operating lever at wide-open throttle as we have discussed in the selection and installation chapter. That chapter also discusses checking and setting fuel levels for bowls with sight plugs. Do it! If your carburetor does not have removable bowls, be sure that the fuel level is correct before proceeding. Remove the air cleaner (temporarily!) so that you can look into the air horn. Have someone else mash the pedal to the floor as you check with a flashlight to make sure that the throttles fully open (not slightly angled). If they're not opening fully, figure out why and fix the problem. Any time you remove and replace the carburetor—check again to ensure that you have a fully opening throttle. It is the easiest thing to overlook and the cause of a lot of lost races or poor times. Any honest racing mechanic will admit that he's been tripped up by a part-opening throttle *at least once.*

For drag racing, air cleaners are generally not used. For any other competition where dust is involved, be sure to use a low-restriction paper-element cleaner, such as a tall open-element AC type and replace it often. Combination paper-element and oiled-foam cleaners should be used for very dusty conditions. If the cleaner is removed, the engine can suck in a lot of abrasive dirt by merely running back down the return road. If the engine has to last, then stop at the turn off and put the air cleaner on—or get the car pushed back to the pits.

The air-cleaner directs the air into the carburetor so the vents work correctly and air gets into the air bleeds correctly. You

Don't drill jets—Jets are carefully machined with a squared-off entry and a precisely maintained exit angle. Drilling a jet will not provide a jet with the same flow characteristics as a genuine Holley machined jet—even though drilling apparently produces the same size opening.

Carrying jets—Store and carry jets in a jet holder such as Holley's punched-out rubber one (Part No. 36-17, holds 42 jets). Or drill and tap a piece of wood, plastic or aluminum with 1/4-32 NF threads and use that. Some tuners use compartmented plastic boxes or even small envelopes. The practice of carrying jets strung onto a wire is strictly a NO-NO because this will very definitely damage the jets.

Use correct tool to install jets—Be sure to use a screwdriver with a wide-enough blade—or use a jet tool—so there will be no danger of burring the slot. Burrs on the edge of the slot can affect main-jet flow.

When you are tuning, watch the results. If the results disagree with your theory—believe the results and invent a new theory!

— Mike Urich

Holley released these main-jet slosh tubes for a Ford NASCAR application with backwards-mounted carburetors. They were important for that particular application. When they are soldered or Loctited into place to hold them in the jets, jet changing can become a real chore.

have probably read a lot of articles which said to be sure to leave the air-cleaner base in place, even if you had to remove the air cleaner for some obscure reason. Lest you think the writers were kidding you, one dyno test series showed a 3 HP loss by removing the air-cleaner base. This little item causes air to flow into the carburetor with less turbulence. It is essential!

Air cleaners also protect against fires caused by starting "belch-backs"—reduce intake noise—and reduce engine wear. Intake noise can be horrendous—even worse than exhaust noise—on an engine turning up a lot of RPM. Air cleaners greatly reduce that noise.

Before leaving the subject of air cleaners, let us remind you to look at the way the stock cleaner is designed before you invest in some flat-topped short cleaner because it looks good. Note that the high-performance cleaner stands high above the carburetor air inlet. Adequate space allows the incoming air to enter correctly and with minimum turbulence. With a flat filter sitting right on top of the air inlet you may get HP loss.

For racing take the sintered-bronze Morraine filters out of the inlets to the fuel bowls. Make sure there is a filter in the line between the pump and the carburetor/s. Also make sure the choke is locked open (*for racing only!*). If you remove the choke (not really necessary) be sure to plug all holes left when you take out the shaft, operating mechanism and fast-idle linkage.

ACCELERATOR-PUMP TUNING

The ability of the engine to come off the line "clean" indicates a pump "shot" adequate for the application. A common complaint—heard again and again—"It won't take the gas." Actually it's the other way around because the engine is not getting *enough* gas and a "bog" is occurring. Two symptoms often appear. The first of these is that the car bogs—then goes. This can be caused by pump-discharge nozzles which are too small so not enough fuel is supplied fast enough. The second symptom is one of the car starting off in seemingly good fashion—then bogging—then going once more. We are talking about drag-race starting here, of course. In this second situation, the pump-discharge nozzles may be correctly sized but the pump is

not big enough to supply sufficient capacity to carry the engine through.

Solving the first problem may mean using larger pump-discharge nozzles—which may subsequently lead to the need for a larger pump, too. More details on tuning discharge-nozzle size appear a few paragraphs later.

In the second case smaller nozzles may be tried in the hope that the existing pump size will then handle the capacity requirement. If that fails, then a larger pump should be tried with the original nozzles.

As a general rule, the more load that the engine sees, the more pump shot that is needed in rate and volume. For instance, if an 1800 RPM stall speed converter is replaced with a 3000 RPM converter in an automatic transmission, the shooter (discharge-nozzle) size can be reduced. The same would be true when replacing a light flywheel with a heavier one. This is because the engine sees less load as the vehicle leaves the line in each instance.

Valve timing affects the pump-shot requirement. Long valve timing (duration) and wide overlap create a need for more pump shot than required with a stock camshaft.

Carburetor size and position also affect pump-shot requirements. More pump shot is needed when the carburetor is mounted a long way from the intake ports, as on a plenum-ram manifold or a center-mounted carburetor on a Corvair or VW. The larger the carburetor flow capacity in relation to engine displacement and RPM—the more need for a sizeable pump shot to cover up the "hole" caused by slamming the throttles wide open. This is especially true with mechanically operated secondaries.

Think about this for a moment and you'll see clearly why Holley warns against converting the vacuum-operated secondary throttles to mechanical operation. There's no pump to cover up the "hole" caused by secondary opening. That's the reason for the "double-pumpers"—to give an adequate pump shot to cover up opening the mechanical secondaries.

If the carburetor is changed from one engine to another or the engine is changed into a different vehicle—or drastic changes are made in the vehicle itself—work may be needed to get the accelerator-pump system to perform as you'd like. The main problem will always be tip-in performance.

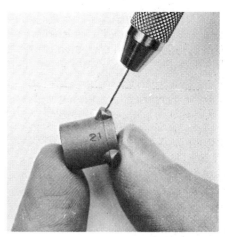

Use a pin vise to hold the drill you use to open up an accelerator-pump shooter. 21 stamped on shooter is original hole size in thousandths of an inch.

This air-cleaner base is used on 1967-71 Corvettes equipped with the L-88 fresh-air hood. To duplicate it, you'd need: Air-cleaner base 6422188, gasket 3919812, screen 3902396, cup retainer 3902394, diffuser 6423906 and wing nut 219281. You'll also need seal 3934165 if you actually use a fresh-air hood.

PERFORMANCE TUNING

Where carburetor calibration changes are discussed in the following sections of this chapter, these are intended FOR COMPETITION ONLY. This means for drag race, off-road and road-race applications—anything where the vehicle is NOT TO BE USED ON STREETS OR HIGHWAYS.

Most Holley carburetors for bolt-on replacement use are calibrated to provide original-equipment specification performance or better when installed without changes. Any modification to the calibrations will probably cause the carburetor/engine combination to be unable to meet exhaust-emission standards/regulations.

Adjustable needle/seat assemblies used on some Holleys with removable fuel bowls are adjusted by loosening the lock screw, then turning the assembly with the nut. This nut is used only for turning the assembly to raise or lower the fuel level. Locking the adjustment is done solely by the lock screw.

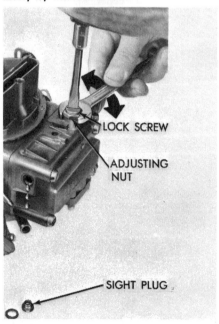

LOCK SCREW

ADJUSTING NUT

SIGHT PLUG

Shooter size tuning is best done by increasing the nozzle diameter (or decreasing it) until a crisp response is obtained when the throttle is "winged" (snapped open) on a free engine (no-load). When crisp response is obtained, increase the nozzle size another 0.002 inch and the combination will probably be drivable for a drag application.

Getting a boat engine to take the throttle can require a much larger shooter size than any other application. 0.040 to 0.042-inch shooters may be needed to overcome the constant load imposed by the propeller.

Pump-shot duration can be timed by shooter size, pump-cam lift and pump capacity. Shooter size also determines the rate at which fuel is fed from the accelerator-pump system during WOT "slams." The override spring on the lever is a safety valve which "gives" (compresses) when the throttle is slammed open. The compressed spring force against the lever operating the diaphragm then establishes delivery pressure in the pump system. Delivery rate then depends on system pressure and shooter size.

The override spring must never be adjusted so that it is coil-bound or has no capability of being compressed. And, do not attempt to replace the spring with a solid bushing in hopes of improving pump action. Either course of action could cause a ruptured pump diaphragm and/or a badly bent pump linkage because gasoline is incompressible. Something has to give or break if the throttle is slammed open and there's no way for the system to absorb the shock.

If the pump cam provides full lift and therefore full travel for the pump diaphragm, the *shape* of the cam is not important for drag-race applications. The cam merely provides a way to compress the override spring so that the spring will cause the pump to deliver its shot into the engine. The *position* of the cam on the throttle shaft is very important for the drag racer. The relation of throttle opening to cam lift is shown in a graph in the chapter on how your carburetor works. This clearly shows that if the throttle is opened very far to provide staging RPM, much of the cam lift is used up so that the pump cannot deliver a full shot.

Ford's Bud Elliott suggests that the racer determine what throttle opening is required to obtain the desired staging RPM. Then the cam should be rotated backward on the shaft until the pump lever again rests on the heel of the cam (no-lift position). You may find that one of the existing holes will line up with a hole in the throttle lever—or it may be necessary to drill another hole in the plastic cam to allow the desired positioning.

Be sure to check that there is NO clearance between the pump actuating lever and the cam. Resetting the throttle to a lower idle speed can move the cam away from the lever—delaying the pump shot. A mere 2° throttle movement should move the pump lever. When readjusting the pump-operating-lever adjusting screw to reestablish contact with the cam, be sure to check that there is 0.015 to 0.020 inch additional travel for the diaphragm lever at WOT between the lever and the adjusting screw.

Various cams are offered for the 2300, 4150/60, 4165 and 4500 carburetors. These are primarily for use in tuning the actuation of the accelerator pump/s on carburetors used for engines being driven with varying throttle openings. This would include street and highway use, road courses and circle-track.

Carburetors which do not use cam actuation of the accelerator pump can sometimes be modified for added capacity by lever/linkage changes to get maximum pump stroke. The pump piston should nearly bottom in its well. Stroke increases may be gained by lifting the piston to a higher starting position but not past the fill slot or pump-well entry. Each carburetor must be examined to see whether the diaphragm bottoms in its housing. If not, minor linkage/cam modifications may allow added travel to the point where the diaphragm is just short of bottoming at full lift.

In a carburetor which does not have removable shooters (passages drilled into body), drilling out the discharge nozzle/s may require disassembling the carburetor. A lead ball or other plug may have to be removed to allow access to the discharge passage.

POWER-VALVE TUNING

You may immediately say, "Ahah—I know just what to do—take it out!" No, regardless of all the magazine articles written to the contrary, *there is seldom ever any real reason to take out the power valve* and replace it with a plug, even though Holley sells plugs for that purpose. The power valve in the secondary is especially important because it allows using smaller main jets. Braking forces do not cause the engine to run excessively rich and there is no tendency for the engine to load up or run rich at part throttle. So many advantages are gained by leaving power valves in place that Holley's engineering staff has gone on record as being strongly against their removal. They point out that the valves have an important purpose or they would not be installed in the first place. If they could be left out, Holley could reduce their manufacturing costs. As discussed in the section on how your carburetor works, the power valve is the "switch" between the mixture ratio for cruising and that required for full power.

Power-valve tuning requires using a vacuum gage. Let's take the example of a car equipped with a camshaft which provides such low manifold vacuum at idle or part-throttle that the power valve opens, giving richer mixtures or perhaps turns on and off due to vacuum fluctuations. In this instance, a power valve which opens at a still lower vacuum should be installed. If a 65 power valve is in the carburetor and the vacuum occasionally drops to 5 inches Hg at idle or part-throttle, install a 40 power valve (opens at 4.0 inches Hg). This ensures that the power valve will not open until the added fuel it provides is needed.

You can see that it is essential to know what the manifold vacuum is at idle. It is also one situation where it is necessary to have a gage which is not highly damped, that is, the needle will have to "jump" to follow the vacuum fluctuations or you won't know how low the vacuum is getting.

Another application which demands using a vacuum gage is racing in a class where carburetion is limited to a certain carburetor type or size which is really too small for the engine which it is feeding. A class demanding the use of a single two-barrel would be typical. Here, the vacuum in the manifold may remain fairly high as the car is driven through the traps at the end of the quarter (that's just one example, of course). The power valve should always have a higher opening point than the

highest manifold vacuum attained during the run, especially at the end. If this is not the case, the power valve will close and the engine will run lean. Disaster will result—usually in the form of a holed piston. Sometimes it will get lean enough to cause "popping" sounds from the exhaust. For example, a carburetor is equipped with a 30 (3.0 inches Hg) power valve. Test shows that manifold vacuum is 4.0 inches Hg through the high-speed portion of the course. Change power valve to a 65 or 85 keeping in mind the previous example so that the power valve does not open at idle because of low manifold vacuum created by camshaft characteristics.

Flat-track racing, especially with super-modified cars—and slaloms—are the only applications which generate G loads high enough to move fuel away from the power-valve inlet so air enters to lean the mixture. If the secondary power valve is removed, the main-jet size must be increased to compensate for the lost area of the power-valve channel restriction. Then flooding through the main jets and out the discharge nozzles will occur during braking or hard stops. The extra-rich mixture during part-throttle operation may cause plug fouling or loading up during "light throttle" use such as warming up or running slowly during caution laps.

TUNING IN THE VEHICLE

A lot of engine-development work related to getting carburetion correct can be done on an engine dynamometer. However, some of the tuning has to be done with the engine installed in the car or boat. In general, it is safe to jet up (richer) one or two jet sizes when moving the engine from the dyno to the chassis or boat.

For tuning—you need a place that is always available whenever you want to use it for tuning purposes. You also need a vacuum gage and a stopwatch. The reason for using the same place is that subtle changes in roads can really throw off your best tuning efforts—unless you always use the same strip. A road can look perfectly level and yet include a substan-

Whistle vent in top metering block provides foam control under hot conditions, but its major purpose is to ensure bowl venting while preventing fuel from spewing out of the bowl vents under extreme cornering, acceleration and braking forces. The first whistle vents were supplied in carburetors used by Ford at LeMans. These vents are available as 59BP-91. Bent-brass baffle with the triangular opening is used in current production—usually only on primary bowls of four-barrels. It serves to keep foam from issuing out of the bowl vents under hot conditions. The perforated baffle in the lower metering block was the first baffle type to be used, but it is no longer used in production. However, it is available as an aftermarket part 57BP-33.

Pulsations in intake manifold affect main-jet size—Main-jet size depends in some degree on the number of cylinders fed by the venturi because fuel flow is affected by air-flow pulsations in the manifold. Pulsing can be extreme at wide-open throttle, where only one or two cylinders are fed by a single venturi with no balance chamber or plenum connecting the cylinder/s with the other cylinders.

tial grade of several percent. You won't be able to tell this with your naked eye. Surveyor's apparatus would be needed to make this kind of judgment.

If you are tuning for top-end performance, and you know the engine RPM at the end of the quarter mile, a lot of tuning can be accomplished with a stopwatch. Start the watch as you accelerate past an RPM point which is two or three thousand below where you want to be at the top end, then stop the watch when you reach the RPM which marks the end of the range in which you are interested. Use the highest gear so you eliminate gear changes. By eliminating these, you get rid of one more variable in the tuning procedure. Stopwatching runs from 3500 RPM to the peak RPM which will be used gives a very accurate indication of whether a change is helping or hurting performance. High gear stretches the time required to pass through the RPM range of interest and keeps away from wheelspin (which often occurs at gear changes, even though it may not be obvious) so you can see the effects of changes in main jet size or pump calibration.

The race course is a poor place to do tuning when competition is going on. There is never enough time to get the combination running exactly right and there is always a lot of confusion in the busy, exciting and emotion-charged atmosphere of the racing situation. The competitor who has to do any more than "fine" tuning or adjustments is literally not ready to race. Further, at a drag race, conditions constantly change so that the times change. For instance, as more rubber is laid down at the starting line, traction "out of the hole" improves and the times get faster . . . without changing the car.

If you can use the drag strip where you ordinarily race, this is an excellent place to tune, especially if the timing devices can be installed and operating to give you instant feedback on how your tuning efforts are succeeding. The idea here is to use the timing devices and the convenience of an unchanging stretch of road to get things right without any time pressures from competitive action.

Standardize your starting procedure, preferably eliminating any standing starts as wheelspin at the line can make a lot of difference in your times—and will con-

fuse your best tuning efforts. If you want to work on your starting techniques, do that separately—when you have the car tuned to your complete satisfaction.

The times or speeds that a car is turning (assuming good, consistent starting techniques) are a good indication of whether the mixture ratio being supplied by the carburetor/s is correct. If a jet change makes the vehicle go faster, the change is probably being made in the correct direction, regardless of what the spark-plug color tells you. As long as a change produces improvement keep making changes in that direction until the speed falls off, then go back to the combination that gave the best time.

Unless the plugs indicate that the mixture is rich, keep making your moves in the richening direction until times start to fall off. Plug color can indicate perfect mixture, even though the engine is getting into a detonation condition at the upper end.

Because the drag-race engine spends such a small portion of the run at a peak-power condition, spark plugs—especially in a quench-type (wedge) combustion chamber—may need to be bone-white for the best times. Don't be concerned about plugs not coloring in drag events so long as the times keep improving as you make one change at a time—and *only one!*

When reading spark plugs, remember that plugs have tolerances, too. One set of plugs may read one way—and another set with the "same heat range" may read differently. If you can't get the plugs to read correctly, try one step colder plugs and then one range hotter. Plugs can often give you a clue as to how the mixture is being distributed to the various cylinders—provided that the engine is in good tune and the compression is the same in all cylinders—and all valves are seating.

STAGGER JETTING

If you are working with an engine which has stagger jetting, then make jet changes up or down in equal increments. If one jet is normally 78 and another an 80 and you are richening by two steps, move up to an 80 in the 78 hole and an 82 in the 80 hole. Keep everything moving equally.

If the engine has "square" jetting (same size in all four holes or same size in primary barrels with a different

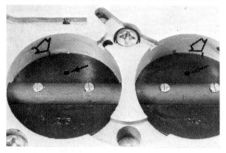

Underside of 4412 version of Holley 2300 500 CFM two-barrel shows use of holes in throttle plates (solid arrows) to keep correct relationship of throttles to the idle-transfer slots (outline arrows). These holes are part of the standard 4412. Note that they are on the same side of the throttle shaft as the transfer slots.

Now You Can Order Hard-to-Find Holley® Parts Direct

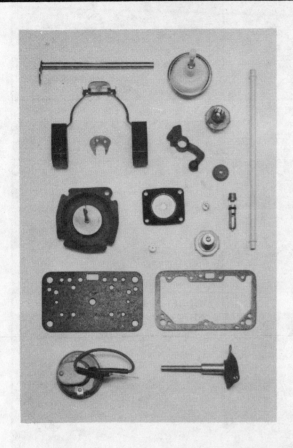

If your local auto store doesn't stock Holley carburetor parts, or other hard-to-find Holley products, you can now order directly from the Holley Customer Service Department in one of two ways:

1. If you have sufficient time, order parts by filling out and returning the order form below. Be sure to include the part no. of your carburetor to assure receiving the correct order form and price information.

2. If you need parts quickly, call **615-859-4924**. If you are able to describe the part accurately, and furnish your carburetor part no., every attempt will be made to handle your order over the phone. If not, the appropriate order form and price information will be sent to you. (Telephone orders accepted with credit card only.)

In either case, the correct carburetor part no. must be provided to Holley. See the reverse side of this sheet to locate the part number, depending on your particular carburetor model.

Prices quoted may vary slightly from published prices at your local auto parts store to allow us to accommodate your order by mail. All prices quoted will be suggested list price, along with any shipping or handling charges.

Note: Warranty claims must be handled at place of purchase.

Cut on dotted line and mail to: Holley Replacement Parts Division, Technical Service Department
601 Space Park North
South Cartwright Street
Goodlettsville, Tennessee 37072

L-30155

- -

Please send me the order form and price information for my carburetor part number referenced below:

NAME _____

ADDRESS _____

CITY _____ STATE _____ ZIP CODE _____

CARBURETOR PART NO. _____

PHONE NO. (____) _____
           Home   Work

Note: Part No. must be included for correct order information.   (Example: LIST 1850-2)

® Registered Trademark of Colt Industries Operating Corp

Holley Carburetor Part No. Location by Model

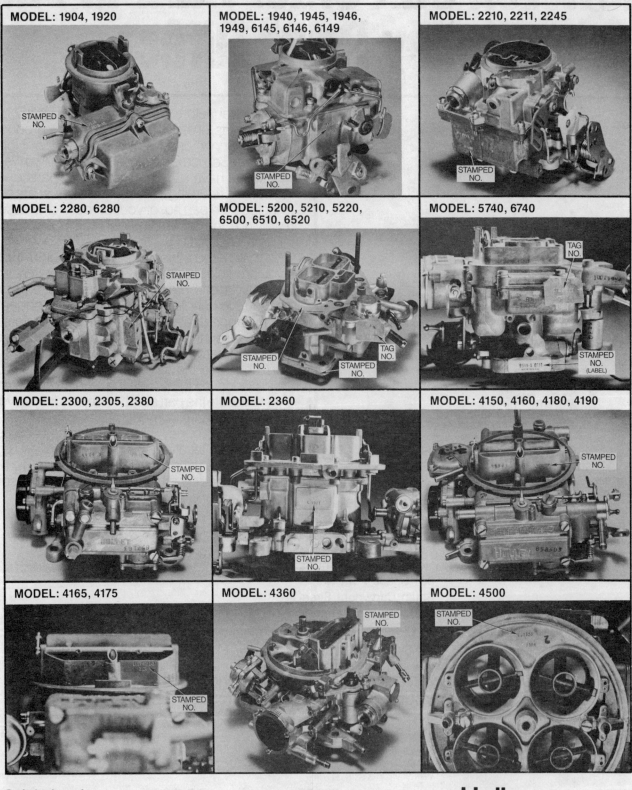

MODEL: 1904, 1920
STAMPED NO.

MODEL: 1940, 1945, 1946, 1949, 6145, 6146, 6149
STAMPED NO.

MODEL: 2210, 2211, 2245
STAMPED NO.

MODEL: 2280, 6280
STAMPED NO.

MODEL: 5200, 5210, 5220, 6500, 6510, 6520
STAMPED NO.
STAMPED NO.
TAG NO.

MODEL: 5740, 6740
TAG NO.
STAMPED NO. (LABEL)

MODEL: 2300, 2305, 2380
STAMPED NO.

MODEL: 2360
STAMPED NO.

MODEL: 4150, 4160, 4180, 4190
STAMPED NO.

MODEL: 4165, 4175
STAMPED NO.

MODEL: 4360
STAMPED NO.

MODEL: 4500
STAMPED NO.

Colt Industries

Holley
Replacement Parts Division

"same size" in the secondaries), make the changes equally for each jet position.

SPECIAL PROCEDURES FOR "WILD" CAMSHAFTS

The next few paragraphs are very specialized and really apply only to the pro racer. Normally such modifications are not required, even with a wild camshaft. So be sure that you try the carburetor in its "box-stock" condition before you proceed with such changes. Holley competition carburetors, such as the various List Numbers of the Model 4150 double pumpers, may be rich enough through the idle and mid ranges to handle a wild cam without changes to idle feeds. However, modifying with a hole in each primary throttle plate is common, especially on the very large engines with wild cams, such as the big-block Chevrolet with the service-package or other long-duration camshaft.

A wild racing camshaft with lots of valve-timing overlap can cause seemingly insurmountable tuning problems. Fortunately solutions are available, although they are not widely known. If you have gone so far down the racing road that you are using a wild "bumpstick," you should not mind the extra effort required to ensure that the engine will idle at a reasonable speed and will not load up the plugs when the car is run slowly—as on warm-up laps, caution laps or running back to the pits after a drag run. This work also makes the engine more controllable in approaching the staging lights at the drags, which in itself could make the effort worthwhile.

The first thing to do is to follow the normal pre-installation procedures of checking the accelerator-pump setting and making sure that the bowl screws are tight. Look at the underside of the carburetor with the throttle lever held against the curb idle stop (not against a fast-idle cam). Note the position of the primary throttle plates in relation to the transfer slots or holes. This relationship has been established by the factory engineers to give the best off-idle performance. The reasons for these slots/holes are described in the idle system explanation earlier in the book.

Check the throttle-plate-to-throttle-bore clearance with a feeler gage or pieces of paper as you hold the throttle

lever against the curb idle stop. Note this clearance in your tuning notebook. Record everything as you proceed, regardless of how good your memory may be.

Install the carburetor on the engine and start the engine. If you have to increase the idle-speed setting to keep the engine running, note how many turns or fractions of a turn open the throttle to this point. Adjust the idle-mixture screws for the best idle. If the mixture screws do not seem to have any effect on the idle quality, note that fact.

Use a responsive (not highly damped) vacuum gage to measure the manifold vacuum at idle. If the engine idles with the manifold vacuum occasionally dropping to a value lower than that required to open the power valve, a power valve which will remain closed at idle must be installed before proceeding with the other changes described in the next paragraphs.

Take the carburetor off of the engine. Turn it over and note where the throttle plates are in relation to the transfer slot. If you can see more than 0.040-inch of the transfer slot between the throttle plate and the base of the carburetor, drill a hole in each primary throttle plate on the same side as the transfer slot. If holes already exist in the throttle plates (some Holleys are equipped with such holes), enlarge these holes. In the case of a 400 to 450 CID engine, the hole size required often works out to be 0.125 to 0.140 inch. Smaller engines require smaller holes. Start with a 1/16-inch drill on your first attempt and then work up in 1/32-inch steps. Before reinstalling the carburetor, reset the idle to provide the same throttle-plate-to-bore clearance which you measured at the beginning of this procedure.

Start the engine. If the engine idles at the desired speed, the holes are the correct size. Too slow an idle indicates that the holes need to be larger; too fast an idle indicates the need for smaller holes. If holes have to be plugged and redrilled, either solder the holes closed or close them with Devcon "F" Aluminum. When

NOTE: Never back off the idle setting screw to allow the throttles to seat against the throttle bores. This can cause sticking or binding and inconsistent idle return. Always maintain the original clearance between the plates and bores.

Throttle/transfer-slot relationship

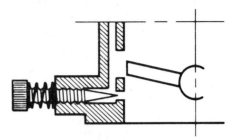

A — Transfer slot in correct relation to the throttle plate. Only a small portion of the slot opens below the plate.

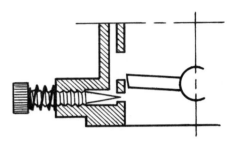

B — Throttle plate closing off slot gives smooth idle with an off-idle flat spot. Manifold vacuum starts transfer-fuel flow too late. Cure requires chamfering throttle-plate underside to expose slot as in A.

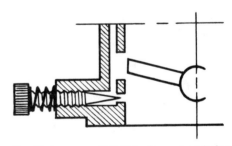

C — If slot opens 0.040 inch or more below throttle, rough idle occurs due to excessive richness and little transfer flow occurs when throttle is opened. A long flat spot results.

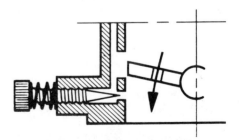

D — Correcting condition in C requires resetting throttle to correct position A and adding hole in throttle as described in text.

Two more idle-fuel-restriction locations. Left is restriction at lower end of idle fuel well. One of these restrictions has been placed on a drill inserted in the actual restriction in the metering block. "Chrysler" type metering block (also used on Model 4500 List 6214 and 6464) at right has idle-fuel restriction in bottom of tube inserted into main well. An extra idle-feed-restriction tube has been placed atop the main well to show relative location. This tube is the idle well. This type of block is identified by the two aluminum spots and the four lead balls on its top.

Idle restrictions may be pressed into the top of the idle well. Brass restriction has been placed on a drill inserted in the actual restriction (arrow).

Wire drill index with drills from 0.0135 to 0.039 inch (Numbers 80 through 61) is an especially useful item for the serious tuner. The pin vise is used to hold the drills. Only finger power is used to twist the drills.

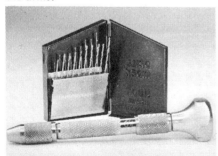

Idle-feed restriction may be a brass tube in the top of the idle well 1, or a brass restriction in the cross channel between the idle well and idle down leg 2.

you have the holes at the correct size (which may require the use of number or letter drills to get the idle where you want it), note the size. Where there are two throttle plates on the primary side, both plates should have the same size hole.

Do the idle mixture screws provide some control of the idle quality, that is, do they cause the engine to run rough as the idle needles are opened? If not, open up the idle-feed restriction approximately 0.002 inch at a time until some control is achieved. Correct control is indicated when the engine runs as smoothly as possible and turning the idle mixture screw either way causes the engine RPM to drop and roughens the idle as the mixture is leaned or richened.

Because you are working here with very small holes (requiring a "wire-drill" set), you must proceed in very small increments. Even a 0.002-inch increase in an idle-feed restriction of 0.028 is an area—and hence, flow—increase of 13%. Wire drills, incidentally, are not used in a power drill, but must be held in a pin vise. Pin vises are available where you buy wire drills, namely at precision tool supply houses, model shops, or through the Sears Precision Tool catalog.

An adequate accelerator pump setup usually eliminates any need to work on the transfer fuel mixture (controlled by the idle fuel feed restriction). The mixture can be checked by opening the throttle with the idle screw until the main system just begins to start, then backing off the screw until it just stops. If richening or leaning with the idle-mixture screw causes RPM drop off, the mixture is correct. Contrariwise, if leaning it causes the RPM to increase, the mixture is too rich and vice versa.

Even if the carburetor is really too big for the engine, try it before making any changes to the idle system. If idle and off-idle performance turn out to be unacceptable, increase the idle-feed restriction slightly.

Make changes in very small increments, remembering that it is all too easy to drill out certain items of the metering block so that there is no easy way to get back to the starting point—and sometimes it requires a new metering block. And, it's easier to take the carburetor off to increase a hole size in the throttle plate/s than it is to solder or epoxy the hole closed and drill another one.

If you are in a time bind and have to go racing before you can make the recommended sequence of tests and adjustments, then a "quick fix" may be of some help. However, it is only part of the cure for this situation in which the throttle is too far open at idle.

Refer back to page 35 where we discussed pump delivery vs. cam type. Pump cam lift is all in by about 20° throttle opening on most of the cams. Thus, a wide throttle opening can use up 40 to 50% of the available pump-shot. On Holley two- and four-barrel carburetors which use a cam for pump actuation, the cam can be rotated to its second position to regain a part of the pump-shot capacity. Only the primary pump cam is moved, of course. Even if you use the cam-relocation "quick fix" you'll still have to take the time to replace the power valve/s with one/s with an opening point occurring at a lower manifold vacuum than you measured with the engine idling.

ROAD-RACE TUNING

Drag racers usually limit carburetion tuning to straightforward and reasonably minor changes: Main-jet size, idle-mixture-screw setting and selection of a power valve which will not be open at idle.

Road-race tuning is something else because these racers go "deeper" in their search for a state of "ultimate tune." We will tell you what these tuners do to get their carburetion "right on," but at the same time we have to warn you that this is not the sort of thing that Mr. Average Mechanic should undertake. This kind of tuning takes experience and the understanding that it is awfully easy to go "too far" in modifying parts which cannot be easily replaced . . . or put back in their original condition.

When tuning involves the idle-feed restriction (IFR) and the power-valve-channel restriction (PVCR) this is getting into carburetor modification which is quite close to carburetor engineering. It requires infinite care and patience—more than most mechanics would ever imagine.

Once the correct carburetor size has been selected, tuning proceeds like this: First the idle is set by turning the mixture screw in until the engine falters and then backing it out 1/8 to 1/4 turn. Next, the free-engine (no-load) RPM is increased slowly to 3,000 RPM to see if there is

any point at which the engine stumbles or misses. The reason for doing this slowly is so that the fuel from the accelerator pump will not confuse the lean condition which is being checked for. If missing or stumbling occurs, this indicates a lean condition. Back out the mixture screw slightly to see whether this helps the condition. If the screw has to be backed out more than 1/2 turn from the best idle setting which was previously established, open the IFR 0.002 inch at a time until there is no stumble or miss to the 3,000 RPM point. NOTE: Read the material on curb-idle throttle-plate positioning related to the transfer slot or holes as detailed for racing camshafts on page 131.

Once the IFR is correctly established for the free-engine tests, the car is road tested for surging at low constant speeds (the mixture is controlled by the IFR from curb idle through 30 MPH steady speed or light-load conditions) and about 14 to 16 inches of manifold vacuum. If the car "feels good" leave it alone. If it surges, additional idle-system fuel may

Power-valve-channel restriction—A large-area PVCR is sometimes used on the secondaries to allow using smaller main jets for best fuel control under severe braking.

Power valve channel restriction (PVCR) changes—When reducing the diameter of the PVCR, the original hole can be closed with lead shot or Devcon F aluminum-base epoxy compound then redrilled to the desired size. Or the PVCR in a metering block can be drilled to allow pressing in idle-feed restrictions obtained from an old metering block. The brass restrictions are pressed in and redrilled to the needed size.

Secondary metering plates. At top is first design used on 3160 and 4160. Center version supplied on some 4160's for Chevrolet had larger well capacity. Bottom plate is aluminum (instead of zinc) for Chrysler 4160's. Numbers identify features: 1—idle-feed restriction, 2—main restriction, 3—main air well, 4—main well, 5—idle well and 6—idle down leg.

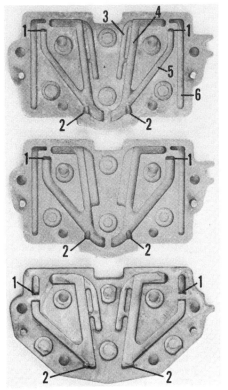
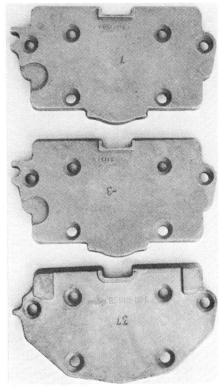

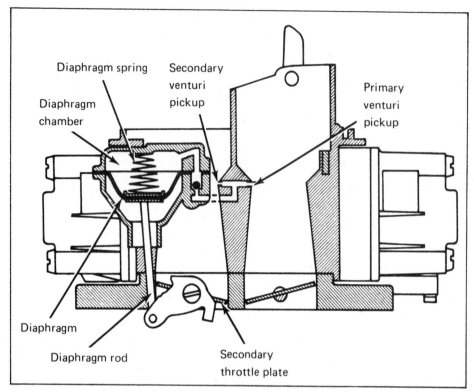

Diaphragm spring

Secondary venturi pickup

Diaphragm chamber

Primary venturi pickup

Diaphragm

Diaphragm rod

Secondary throttle plate

cure the condition. This is obtained by increasing the size of the IFR.

Next, starting at 30 MPH, a series of "crowds" are made with the driver "crowding" a certain manifold vacuum as observed on a gage. These crowds are light accelerations made while keeping the manifold vacuum first at 12 inches, then at 10 inches, then at 8 inches, and so on—to the point of power-valve opening. If there is surging during these crowds, attempts to improve the condition are made by varying the size of the main jet. If there is a lean surge, the primary jet size is increased until acceptable drivability is obtained. Incidentally, a rich condition is not usually found—except in some high-performance carburetors. Keep track of the main-jet size which provides the desired drivability because you will have to know what that size is at the conclustion of the next test.

Next, wide-open-throttle tests are made from 20 to 80 MPH in high gear. The main jet size is increased or decreased to get the best time, as determined by stopwatch. If the jet size turns out to be larger than that established for drivability in the previous test, the power-valve-channel restriction should be increased in area to compensate for the difference in area between the two main-jet sizes. A drill size which provides the desired area increase is used to open

the PVCR. Then the carburetor is reassembled, installed and the car is retested to ensure that the changes have been made correctly. If a smaller PVCR is required, this is more difficult to accomplish because many PVCR's are merely holes which are drilled through the metering block from the power-valve area to the main channel. In general, reducing the PVCR area is more difficult than increasing it because the hole may have to be plugged and drilled to a smaller size.

The accelerator pump tuning has already been described for street and drags, but there may be instances where a road racer will have to modify the pump-cam phasing on the secondary (assuming a double-pumper carburetor is used) to get correct throttle response coming off of a turn which has required slowing down to the point where the secondaries are just starting to open or are slightly open. This kind of tuning can only be done for the individual track/course situation and is only mentioned here as one of the items to be considered.

Remember that conditions vary from hour to hour and day to day. Your tuning results will only relate to all things accomplished on the same day. Always take the time to rerun your baseline or best times to make sure that this has not changed because of variables such as engine condi-

tion, plug condition, tires, atmospheric pressure, etc.

TUNING VACUUM SECONDARIES

Vacuum-operated secondary throttles are initially opened by a signal (vacuum applied to the diaphragm) from the primary venturi as air flow through the venturi is increased with engine RPM. This signal is bled off by a second, smaller hole into the secondary venturi (called a "killer bleed") so that the signal being applied to the diaphragm increases smoothly. When primary air flow gets high enough to create a signal which overcomes the diaphragm spring, the secondary throttles begin to open . . . even though some of the signal bleeds off through the hole into the secondary.

As air flow starts to increase through the secondary venturi, the secondary bleed hole adds signal to that from the primary. This signal (1) helps to open the secondaries and (2) eliminates "hunting," "flutter" or unstable operation of the secondaries near the opening point.

The reason for the vacuum-operated secondaries, of course, is to ensure that the secondaries will not open ahead of the time when the air flow is needed. The engine is never over-carbureted even if the driver punches throttle wide open at low speed. The engine goes right on operating on the two primary barrels for good low-end response until the engine builds up RPM, creating sufficient primary air flow so the diaphragm opens the secondaries when the engine can use the extra capacity of the secondary barrels.

Vacuum-operated secondaries will open at different full-throttle RPM, depending on the spring which is installed behind the diaphragm which operates the secondary throttles. A larger engine opens the secondaries sooner than a smaller displacement engine because the larger engine generates higher air flow.

The opening point and rate of opening for the secondary throttles can be tailored to the engine and application by using one of several diaphragm springs. This is really the only item to consider changing when working with vacuum-actuated secondaries. This spring counterbalances the vacuum signal applied from the venturis, holding the secondary throttles closed against the

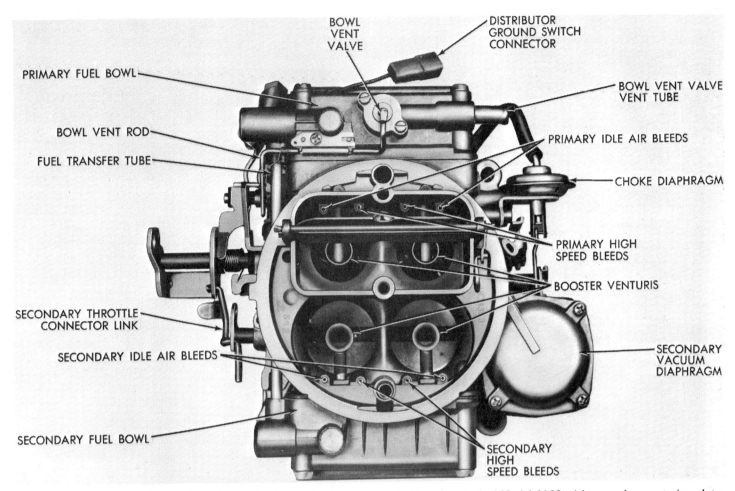

PRIMARY FUEL BOWL

BOWL VENT ROD

FUEL TRANSFER TUBE

SECONDARY THROTTLE
CONNECTOR LINK

SECONDARY IDLE AIR BLEEDS

SECONDARY FUEL BOWL

BOWL
VENT
VALVE

DISTRIBUTOR
GROUND SWITCH
CONNECTOR

BOWL VENT VALVE
VENT TUBE

PRIMARY IDLE AIR BLEEDS

CHOKE DIAPHRAGM

PRIMARY HIGH
SPEED BLEEDS

BOOSTER VENTURIS

SECONDARY
VACUUM
DIAPHRAGM

SECONDARY
HIGH
SPEED BLEEDS

This Holley illustration refers to the main air bleeds as "high-speed bleeds." Photo is a typical Model 4160 with secondary metering plate and a primary metering block. The float bowls on this example are side-hung types with non-adjustable needle/seat assemblies. This unit has vacuum-actuated secondary throttles.

curb-idle stop until the diaphragm receives sufficient vacuum force to compress the spring and open the throttles. Atmospheric pressure against the offset secondary throttles assists in keeping them closed. Inconsistent idling can be caused by using a clipped spring or a non-standard spring behind the diaphragm.

It is always best to start your tuning efforts with the standard spring and then use springs from kit 85BP-3185 to accomplish any changes. The secondary section of the chapter on how your carburetor works provides a graph of spring height versus load. The yellow spring allows the secondaries to open quickest because this spring has the lowest load at a given height. Carburetors are usually supplied with a green, purple or red spring.

If you have the triple two-barrel setup with vacuum-operated end carburetors, you'll have to use the blue spring that

comes in the carburetors as no others are offered. The blue spring is not shown on the spring-action graph.

A number of "tricks" have been touted as important modifications for vacuum-secondary Holleys. Most are almost—if not completely—useless! The first thing that many mechanics do is to take out the check ball because it is an easy thing to do even if they don't understand what really happens. A bleed groove in the ball seat allows the signal from the venturis to build at a controlled rate so that the secondaries do not open suddenly. When the throttles are closed, the ball unseats to bleed the vacuum signal from the diaphragm immediately, thereby allowing the secondaries to close quickly. Closing the throttles quickly is more important than allowing them to open suddenly in nearly every instance. Taking the ball out of the circuit allows you to

"feel" the secondaries open. Many have mistakenly interpreted what they feel as an increase in acceleration. Surprise! What they are really feeling is a bog or sag in the acceleration curve.

The recommendation most often heard and one which has destroyed the drivability of more carburetor/engine combinations than any other is that of changing diaphragm carburetors to mechanical operation of the secondaries. A screw is placed in the secondary lever and the diaphragm or its spring is removed. Secondary operation is then mechanically controlled by the primary throttle shaft. Those who "push" this conversion overlook the fact that the carburetor now requires more care in driving. No pump is available to cover up the "hole" or bog created when throttles are opened suddenly without regard to engine RPM. This is especially true if the secondaries are operated simultaneously with the primaries (1:1 throttle action). If you want mechanical secondaries for a racing application, then you really need another carburetor—one with two accelerator pumps or a center-shooter type.

The best recommendation we can make is to let the engine open the secondaries. Change the opening point by swapping springs if you like, but keep the original spring so you can reinstall it if you don't find help by using different ones. Watch the acceleration times closely so you are not misled by "the feel in the seat of your pants."

So you won't spend a lot of time fighting a problem that has been overlooked by many—check the secondary throttle operation by hand as you hold the primary throttle wide open. The throttle shaft should move easily against the resistance of the diaphragm spring and close easily against the stop. Binding can be caused by deposits on the shaft if the secondaries have not been used. And, it is quite common for the shaft to bind due to incorrect and uneven tightening of the carburetor attachment nuts/capscrews. This problem arises most often when a soft gasket or a gasket pack has been used.

Air cleaner affects diaphragm-secondary operation—Unless the engine was originally equipped with a very low-restriction air cleaner, a weaker diaphragm spring will be required to get the same secondary opening point when the air cleaner is removed. This is because the air cleaner restricts the flow, providing a higher vacuum signal to operate the diaphragm at a lower RPM than when the air cleaner is off. This is typical of diaphragm-equipped carburetors from early Fords and Chevrolets with the single-snorkle air cleaner. A yellow spring is a good choice when starting tuning. If it opens the secondaries too quickly, other springs can be used to get slower action until the desired opening point has been found.

OFF-ROAD TIPS

With the increasing popularity of off-road and recreational vehicles, many have approached us with questions about optimizing carburetor performance for these applications. Two basic problems are involved: (1) angularity and (2) vibration.

Angularity—Visualize the problem as you tip the carburetor. Consider how the change in the fuel level changes its relationship to the main jets, discharge nozzle and fuel-bowl vents when the carburetor is tipped. Note that the angle can affect the height of the discharge nozzle in relation to the fuel level. In some attitudes, a 40° tilt may cause fuel to drip or spillover from the discharge nozzle. Other than using a military-vehicle carburetor, or the Buggy Spray (List 6113), a Model 1940 or older Holley center-float carburetors (1901, 2140 and 4000), there is one "fix." Drop the fuel level approximately 1/16 inch. The problems with lowering the fuel level are: The power valve and main jets are uncovered sooner and nozzle "lag" may create a noticeable off-idle bog. Part of this lag can be offset with increased pump (either by opening the discharge nozzle/s or "shooters," or using a larger accelerator pump—or both). Some lag or "bog" is almost inevitable, especially if the carburetor is too large for the engine and RPM being used. The problem is minimized by using a carburetor no larger than absolutely required for the peak RPM to be expected in actual driving. You very seldom need large carburetors for off-road use.

Vibration—The float takes a real pounding in off-road applications. The bumper spring under the float must be selected so that it will assist in damping the wild gyrations of the float as it is vibrated. In general, the spring should be strong enough so that it will just allow the float to drop of its own weight.

In some instances this could require using two bumper springs wound together, or a spring from another carburetor. Two springs are offered for the side-inlet fuel bowls (4150/60, 4165, 2300). 38R-757 is the standard one and it has a plain steel finish. 38R-803 is a bit stronger and is colored blue.

The bumper spring takes care of some

of the downward movement, but snubbing the upward travel requires a spring-loaded needle and seat assembly to avoid ramming the needle into the seat with great force. A spring-loaded needle and seat assembly for 4150/60, 4165, 2300 and 4500 carburetors is the 18BP-268AS. This is a 0.082-inch inlet valve. Holley makes side-inlet fuel-bowl assemblies incorporating bumper springs and spring-loaded inlet needles. Part number is 85R-4424. It uses the same 0.082-inch inlet valve.

Vent or Pitot Tubes—These should be extended upward as high as possible to handle sloshing fuel which can come out of the vent tubes under severe braking and/or bouncing. Off-road cars with remotely mounted air cleaners are sometimes equipped with small hoses to carry the vents all the way up into the air cleaner.

Throttle Linkage—The use of a cable drive is often recommended for off-road installations to transmit less vibration to the driver's foot. Many stock automobiles are now equipped with such arrangements and it is usually easy to obtain a cable throttle linkage with everything made to fit the job at hand. Hydraulic linkages are also popular, but it is quite easy to overstress the throttle levers with such hook-ups. The hydraulic linkage should be installed so that the full travel of the actuating cylinder provides wide-open throttle. In some instances, the cylinder mounting has to be carefully thought out so that there is not an over-center condition during throttle operation.

Air Cleaner—The best possible air cleaner is one which completely encloses the carburetor, with the throttle operated by a cable. The cable housing can be sealed where it enters the baseplate for the air cleaner. More details on off-road air cleaners are in H. P. Books volume, *Baja Prepping VW Sedans & Dune Buggies.*

Carburetor & Fuel Economy

TEST METHODS

Fuel economy and data about it have received increasing attention in recent years. So many different numbers are quoted and so many disclaimers made that a guy can get downright confused. I'd like to talk about the more commonly quoted numbers and clear up some of the confusion. Hope I don't add to it.

EPA ECONOMY NUMBERS

The EPA issues two different kinds of economy numbers, city and highway. First, I'll try to explain the EPA or Environmental Protection Agency. This Federal agency—among other things—sets automotive emission standards and test procedures and conducts vehicle-certification tests to assure itself that the manufacturers are in compliance.

EPA economy numbers are derived from data taken from the CVS type test described in the emission chapter. Exhaust gases are measured in actual grams per mile during test runs over a simulated 7.5-mile route All compounds containing carbon, namely CO, CO_2 and HC, are so recorded. Because the relationship between carbon and hydrogen atoms in the fuel is known, the weight of the carbon in the fuel can be calculated in grams per gallon by using the molecular weights of the two. Test results give each of the carbon compounds in grams per mile. Using the molecular weights of carbon, hydrogen and oxygen, total weight of emitted carbon can be calculated in grams per mile. Dividing grams per gallon by grams per mile, they end up with miles per gallon. How about that! The EPA released this kind of data for the first time with the 1974 car model certification data. I can't understand why they didn't simply measure the fuel consumed during the test and divide it into 7.5 miles. My engineering group tried it on one test and got numbers 9% higher than those calculated by the EPA method.

A lot of criticism was directed at the EPA during 1974 because the numbers were quite low. This could have been

Holley's 1976 economy-carburetor line-up. From lower left and going clockwise are Models 1920, 2300, 4360, 5210, 2210, 1940. Center carburetor is a 5200.

predicted because this cycle represents city driving. So for 1975 the numbers just described were designated as *city economy* and the EPA designed a *highway fuel-economy procedure* and began releasing these numbers, too. The highway economy test and calculations are conducted just like the city tests except testing is done with a completely warmed-up engine and the driving cycle has a higher average speed with fewer accelerations and decelerations. Trip distance is 10.24 miles

SAE ECONOMY NUMBERS

The automotive industry traditionally measured fuel economy by driving the vehicle through a prescribed course and measuring actual fuel consumption. Every manufacturer has a slightly different procedure, but all give city, highway and constant-speed fuel economy.

In 1974, the SAE (Society of Auto-

motive Engineers) designed a standard fuel economy test consisting of a driving-event schedule for three distinct cycles. The city portion has 15.6 MPH average with seven stops. A suburban cycle runs at a 41.1 MPH average with two stops. The non-stop interstate cycle averages 55 MPH.

In November, 1974 Union Oil Company and NASCAR ran SAE economy tests at Daytona using 58 1975 vehicles of all makes. Results were reported in the November 18th and 25th, 1974 issues of Automotive News. They used the same kinds of vehicles used by the EPA in their certification and fuel economy reporting so a direct comparision could be made. All vehicles were emission-tested and only those that passed were used. City-economy results averaged 18% lower than EPA city numbers; interstate numbers averaged 8% lower than EPA highway results.

OTHER TEST METHODS

Another traditional way of looking at fuel economy is the constant-speed or road-load-hook method. The vehicle is driven at constant speeds, usually from 20 to 60 MPH while fuel consumption is measured by one of many metering devices. Distance is measured with a fifth wheel trailing behind the vehicle. Economy is calculated in miles per gallon. Main jets are changed in the carburetor and the tests rerun. This allows plotting fuel economy verses main-jet size at each speed for that particular vehicle package. Of course, this same procedure is also used to evaluate the effects of other engine variables such as spark advance.

Steady-state economy can also be run on the dynamometer. Such tests vary among manufacturers. One method is to determine engine load for various vehicle speeds either by actual test or empirically. Engine test speeds are determined by multiplying vehicle speeds by the N/V ratio for the vehicle: Simply, engine RPM divided by vehicle speed in MPH. The engine is maintained at a given load and speed on the dynamometer. Fuel/air ratio is changed either by affecting fuel-bowl pressure or by changing main jets. Throttles are opened or closed to maintain power. The test is repeated at other speeds and loads. Data is plotted as brake specific fuel consumption, pounds per HP hour versus manifold vacuum. These plots not only allow selecting the most economic fuel flow at each speed, but also permit engine-to-engine comparisons.

Well, there you are! Hope I didn't confuse you more. I feel the EPA method is probably the best simply because so many factors like temperature and vehicle load are closely controlled. Therefore, results, while not having great significance on an absolute basis, seem very repeatable, yielding a good benchmark for year-to-year and vehicle-to-vehicle comparisons.

CARBURETOR EFFECTS

Liquid fuel doesn't burn very well, so the object is to get it vaporized or at least atomized into the smallest possible droplets. The carburetor can help do this by supplying strong metering signals. Small venturi sections and efficient booster designs accomplish this. Proper preparation of the fuel and air emulsion in the main well also affects atomization. Heat supplied either to the inlet air or in a hot spot in the intake manifold benefits to vaporization but has a negative effect on power because it lowers charge density. Once you've got the fuel in suspension, good manifold design helps keep it that way.

If leaning out at any load and speed can be prevented, the overall mixture can be kept less rich and economy will improve. For example, the idle system is controlled by manifold vacuum so at a low-speed, low-manifold-vacuum condition the idle-metering signal is lowered and there is a leaning effect because that system's fuel feeding almost completely stops. The best way to overcome this problem is with an efficient, early-operating main system. An ideal carburetor allows the programming of the correct fuel/air ratio at any load, speed and transient condition.

We are constantly confronted with the question; which gives better fuel economy, two-barrel or a four-barrel carburetor? The question is not always easy to answer. In Chapter 2 we showed how staged carburetors expand the effective metering range. It all depends on how we choose to use that expanded metering range. When the primary size of a staged carburetor is kept the same as a non-staged carburetor it replaces, you can expect about the same fuel economy. If you split this metering range by using a smaller primary size you can expect to run leaner mixtures and hence get better fuel economy. The greater total flow still gives us some power advantage. Earlier staged carburetors tended to follow the former route, while in recent years the trend has been toward the latter. Examples are the Model 4165, 4175 and 4360 four-barrels and Model 5200 and 5210 staged two-barrels.

If you drive these carburetors the same way you drive a non-staged one economy is usually improved. If extra performance is there, there is a strong temptation to use it and that consumes fuel.

As stated in Chapter 1, good cylinder-to-cylinder fuel/air distribution is essential for improved fuel economy. If all cylinders are running at nearly the same F/A ratio it logically follows that the overall mixture can be kept leaner. Vaporization, atomization and good intake-manifold design are critical in setting good cylinder-to-cylinder distribution. In fact, to optimize distribution, the carburetor and manifold must be treated as a system.

Another item used in recent years to aid fuel economy is staged or gradient power valves. These valves allow increasing the fuel/air ratio moderately in the higher end of the part-throttle range and bringing in full-power enrichment as late as possible. Most of the new Holley carburetors and some of the old ones have this capability. Operation of these valves is explained in detail in Chapter 2.

OTHER ENGINE FACTORS

Spark timing greatly affects fuel economy. Unfortunately, most engine-modification emission packages used in recent years include spark retard to control unburned hydrocarbons and oxides of nitrogen. Engines are made less efficient so exhaust temperatures can be higher to reduce hydrocarbons and peak temperatures in the cylinders lower to reduce oxides of nitrogen. This reduced efficiency also reduces fuel economy. More energy goes out in the form of heat and is wasted. It causes higher engine-compartment temperatures which deteriorate hoses and various other rubber and plastic parts.

Higher compression ratios increase engine efficiency. There has been a trend toward lowered ratios in recent years for the reasons stated above as well as lowering the octane requirement so unleaded fuels can be used. Unfortunately this also hurts fuel economy.

Exhaust-gas recirculation (EGR) systems take a percentage of the exhaust gas and reroute it through the engine. As discussed in the emission chapter this helps control oxides of nitrogen. Because burning speed is reduced, efficiency and fuel economy drop. The dilution of the intake charge requires a richer mixture and the throttle has to be opened further because of loss of power. These items also adversely affect fuel economy.

Another dramatically changed part on emission-controlled engines is the camshaft. By adding overlap—the time when both intake and exhaust valves are open—the cam provides added EGR through the manifold. Then the gear ratio for the rear end is established with a low numerical ratio so the engine can never operate in an RPM range where it can effectively use this added overlap for speed or power. The result is a further reduction in fuel economy.

There *is* some good news. Most manufacturers have gone to high intensity electronic ignition systems in an attempt to burn leaner mixtures, aiding economy.

The addition of the catalytic converter also aids fuel economy because it treats emissions in the exhaust system after the burning process is over, allowing the car makers to put some tune and efficiency back into the basic engine.

Many other items affect fuel economy. To name a few:

1. Drive train and rear axle ratio
2. Vehicle weight
3. Wind resistance and aerodynamics
4. Accessory loads
5. Engine size

With all of this emphasis on fuel economy, early in 1975 Holley began looking at the idea of replacement carburetors to improve fuel economy. Many replacement carburetors were designed some years ago. Ongoing development programs indicated areas of potential improvement. We studied the existing designs to figure out how to incorporate newly found economy improvements.

A goal of 5–10% improvement was set, using EPA test procedures as a basis. Early results were encouraging so we decided to add three other objectives: Reduced cost, easier installation and wider application. So the same carburetors had to meet emission standards for the recommended applications. Also, they would fit more vehicles.

TEST PROCEDURES

EPA highway tests were run according to federal procedures. EPA city economy tests were slightly altered. Instead of starting with a cold vehicle that had "soaked" or sat for 10 hours, testing was done with a warmed-up vehicle. The reason was that very small gains were being sought and sometimes the day-to-day and test-site variations were greater than the improvement itself. To eliminate these variations the vehicle was placed on the emission rolls and warmed up. Then city and highway economy were run with a given carburetor combination. Leaving the vehicle on the rolls, the carburetor was changed and the tests rerun immediately. As a result absolute city-economy numbers are a little higher than EPA reported ones but the object was to obtain repeatable comparative results. This method also allowed getting a lot of data in a short period.

Actual exhaust-emission tests to verify compliance were run according to federal procedures from a cold start.

To get a feel as to how the EPA results related to the real world, several carburetors were evaluated on an actual road test: A 200-mile city and highway trip through Southeastern Michigan. No sophisticated equipment was used. Vehicle tanks were filled before and after every test at exactly the same spot. Corrected odometer readings were used for the mileage. A control car was sent along on every test to correct for day-to-day variations.

ECONOMY CARBURETORS

There are seven different models with a total of about 50 list or part numbers in our first group of economy carburetors introduced early in 1976: Models 1920, 1940, 2210, 2300, 5200, 5210 and 4360. All are described in detail elsewhere in this book, so I'll limit my discussion to applications, new features and test results. All test data is presented as a percentage compared to the baseline or original carburetor on the vehicle. The model 4360 is not covered here because its economy features are discussed in its own section. It is the only new carburetor in the group.

MODEL 1920

This single-barrel model is used on Chrysler six-cylinder engines from 1960-73 and AMC from 1964-69. Added features include a new booster venturi for better atomization, a high-velocity idle system and a two-stage power valve. High-velocity idle is achieved by opening the system up. Larger metering ports and transfer slots cause more air to be mixed into the idle passages, resulting in higher velocities and better atomization. Tests indicated 7.1% gain in city economy, 9.5% gain in highway and 15% gain in actual road tests.

Looking down the throat of the 1920 you can see the new nozzle. Vanes help move the fuel around for better mixing.

Two-stage power valve for the 1920. Plastic tangs are different lengths. As the stem acts on the valve the two valves open at different points, performing a stepped function.

Two 1920's demonstrate the opened-up idle systems. Economy unit at left has a much wider transfer slot (arrows).

The new Model 2300 economy carburetor.

MODEL 1940

This single-barrel model is for six-cylinder Fords 1962-73, AMC 1970-74, Chevrolet, Buick and Oldsmobile 1968-74 and Pontiac 1970-71.

The booster venturi in these carburetors is undercut slightly to aid in fuel dispersion. A high-velocity idle system is used with staged power valves on some applications. Average gains for these applications are 11.6% city, 8.2% highway and 6.2% on the road test.

MODEL 2210

Applications for this two-barrel are Chrysler V8's 1963-74 and Pontiac, Chevrolet and Buick 1968-74. Special features include the high-velocity idle system and a two-stage power valve. Average gains in this case were 2.9% city and 5.4% highway. No road tests were run with these vehicles.

MODEL 2300

This two-barrel carburetor serves V8 Fords 1961-74 and AMC 1968-74.

Because this carburetor is the oldest of the group, we had to make the greatest number of changes on it. Some of these were to make the carburetor more of a bolt-on replacement. This made installation easier. First, a new fuel bowl was added with the fuel inlet to match the fuel line of the vehicles. This resulted in a float and inlet system that is a mirror image of the old one. The choke housing was redesigned to hook up directly to the existing choke tube. The throttle body was redesigned to incorporate a universal PCV tube and eliminate the need for a spacer in some applications.

Economy-oriented changes include new discharge nozzles, high-velocity idle

Fuel bowls of the Model 2300. Old on the left, new on the right. New one hooks up directly to the original fuel line.

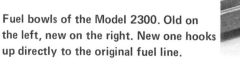

Traditional booster venturis are shown on the left. New improved vaned types are on the right. The new larger booster reduces air-flow capacity slightly.

Old choke housing is on left, new is at right. New one makes a direct connection to vehicle plumbing.

systems and a two-stage power valve. Economy gains are 5.8% city and 6.5% highway.

MODELS 5200 and 5210

The Model 5200 is used on Pinto and Capri vehicles from 1971-74. The 5210 is for the 1973 and 74 Vega. Because these staged two-barrel carburetors are relatively new, little could be done to improve them. Double booster venturis are used on the primary side for stronger signals and better atomization. Economy gains are 4.9% city and 3% highway for the Ford 2300cc engine; 5.4% city and 6.3% highway for the 140 CID Vega engine.

SUMMARY

This economy line will be expanded to include other models and later coverage.

While these gains may seem relatively small, they are significant when you consider most of the carburetors were pretty good to begin with. The carburetor is just one of many factors affecting fuel economy.

Conservation of energy and gasoline shortages have brought fuel economy into sharper focus than ever before. Carburetor and engine engineers have long been concerned with the factors affecting fuel consumption. Cylinder-to-cylinder mixture distribution and good vaporization are a couple of these factors. Overall fuel/air ratio is another one. There are definite limits of fuel/air ratio that will support combustion. Most current carburetors are calibrated very near the lean limit in the part-throttle or economy range. There are some cases where minor improvements can be made by recalibration, but the old fairy tale about a carburetor that will give you 50 MPG on a 450 CID engine in a 5,000-pound vehicle is just that—a fairy tale. There are things *you* can do to get the maximum fuel economy out of your vehicle. Here are some of them:

1. Keep your engine in tune—A fouled spark plug can have drastic effects on fuel economy and power. If the plug is completely misfiring, that particular cylinder is pumping all of its charge right out the exhaust pipe.

2. Ignition timing—Make sure the basic distributor timing is correctly set and all vacuum lines are secure and intact. Engine efficiency drops off with spark retard and that is money out of *your* pocket.

3. Inspect the air-cleaner element—A dirty filter causes greater resistance to air flow, thereby creating stronger metering signals and richer mixtures in some carburetors. This added restriction also reduces the horsepower capability of your engine.

4. Check the choke—The choke plate in the carburetor should be in a vertical position when the engine has reached normal operating temperature. Inspection is quick and easy after removing the air cleaner. Consult the troubleshooting section if a problem exists.

5. Increase tire pressure—This will result in a slightly harsher ride, but rolling resistance will be lower, resulting in

New throttle body on the right is much higher, allowing the use of a universal PCV tube and eliminating parts.

Model 2300 staged power valve installed in the metering block.

better economy. Your tires will last longer, too.

6. Avoid unnecessary hard accelerations—By sensing manifold vacuum, the carburetor automatically increases the F/A ratio as load increases. Manifold vacuum decreases as load is increased. This idea is to keep manifold vacuum as high as possible. Mixtures in the power range are 10—15% richer than the economy range. If you have a vacuum gauge in your vehicle, keep vacuum above 6 inches of mercury.

7. Plan your trips—One trip with several short stops is much better than a lot of short trips several hours apart. The object here is to keep the automatic choke—with its richer mixtures—from coming back on. Also, frictional horsepower or the power required just to move all the engine parts is greater with a cold engine.

8. Watch your speed—Considerably more horsepower is required to move a vehicle at 70 MPH than at 50 MPH because wind resistance increases with speed.

9. Keep your speed constant—Don't let your speed drift up and down. Acceleration requires energy. Also, the accelerator pump gives a little squirt every time you move the throttle.

Another H. P. author, Doug Roe, has written a whole book on the subject of economy. It is called *The Whole Truth About Economy Driving.* If you drive a car, this book will help you save money. He tells you everything you must know to make practical use of the driving and tuning techniques he has been using for years to win national fuel-economy runs time after time—including the 1974-75 Union 76 Fuel Economy Tests in Florida and California. Tuning tips and engine-modification information tell and show you how to get extra miles from every gallon. On top of all this, Roe exposes rip-off economy gadgets. He even includes straightforward advice and info to help you choose the right car—balancing economy and practicality for the driving you do. What he says is interesting, authentic and readable.

Carburetor & Emissions

Air pollution is a worldwide problem. This atmospheric condition is not limited to urban or industrial areas—it is now almost impossible to get away from pollution, regardless of where you might be on this earth. Growth in population and fuel use has been a major contributing factor. The magnitude of the problem—and the concern of the public—is manifested in state and federal legislation which has been enacted to regulate the amount of contaminants which can be put into the atmosphere by automobiles. Standards for industrial emissions have also been set, but the major focus and enforcement has been in the area of new automobile manufacturers and makers of aftermarket (replacement) parts which directly affect emissions.

California, specifically Los Angeles and San Francisco areas, has the greatest smog problem and hence this state has led in exhaust-emission-control regulations.

So, what is *smog*, anyhow?

It is a simple term for a complex happening. When unburned hydrocarbons (HC) and oxides of nitrogen (NO_x) combine in the atmosphere and are acted upon by sunlight, complicated chemical reactions occur to produce *photochemical smog*. Both areas mentioned have all of the necessary ingredients: HC + NO_x + sunlight. These are aided in combining by the inversion layer over these areas—a dense layer of the atmosphere which prevents the escape of the ingredients into the upper atmosphere where they might be swept away by winds. The inversion layer is often likened to a lid on a pot and this "lid" holds the ingredients right there so that the reaction has plenty of time to take place—often several days at a time.

Smog causes nose, throat and eye irritations. And, like carbon monoxide (CO), is extremely harmful to animal and plant life (including trees). Smog also cause deterioration of some plastics, paint, and the rubber in tires, seals, weatherstripping and windshield-wiper blades.

Holley Emission Test Laboratory. These two installations are duplicates and run completely independent of each other. Vehicles are rolled onto the chassis dynamometers and load is controlled as dictated by test procedures. Test schedule is driven by trained technicians who maintain the speed vs. time sequence indicated by a moving chart. Analyzers are to the rear. Plastic sample bags on extreme opposite sides of the room hold the exhaust-gas sample. Samples are analyzed for mass value of each constituent.

There are three major types of vehicle emissions to consider:
1. Crankcase
2. Exhaust
3. Evaporative

CRANKCASE EMISSIONS

The first emission-control device—required by California in 1961—was subsequently required nationwide. It was a metering valve plumbed between the crankcase and intake manifold. Before 1961, all engines were vented into the atmosphere through a road-draft crankcase ventilation tube. The vent spewed unburned hydrocarbons and carbon monoxide into the atmosphere. Although the first systems pulled air through the wire mesh of the oil filler cap, the systems were rapidly changed to a closed, or

Positive-Crankcase-Ventilation (PCV) system. By 1965, this was the standard type. Air from the air cleaner or a filter enters the crankcase, usually through a tube connected to an oil filler cap in one of the rocker-arm covers. Air flows through the engine, and exits—usually on the opposite side or end of the engine—through a tube containing a PCV valve. This valve is connected into the intake manifold. The valve has a spring-loaded large area poppet and a fixed restriction. At periods of high manifold vacuum—as at idle or on the overrun—flow is only through the restriction, so very little flow occurs. As manifold vacuum drops, the spring in the valve overcomes manifold pressure so that PCV flow increases with engine speed. This closed system also

Federal and California Automotive Exhaust Emissions Standards
(Passenger-car and light-duty vehicles to 6000 lbs. GVW)

Year	HC	CO	NOx
California Cycle			
1968-69	275 ppm	1.5%	—
	(3.2 gm/mi)*	(33 gm/mi)*	—
1970	180 ppm	1.0%	—
	(2.2 gm/mi)*	(23 gm/mi)*	—
1971	2.2 gm/mi	23 gm/mi	4 gm/mi C
Federal Test Procedure			
1971	4.6 gm/mi	47 gm/mi	—
1972	3.4 gm/mi	39 gm/mi	3.2 gm/mi C
1973-74	3.4 gm/mi	39 gm/mi	3.0 gm/mi
1975-76	1.5 gm/mi	15 gm/mi	3.1 gm/mi
	0.9 gm/mi C	9 gm/mi C	2.5 gm/mi C
1977	1.5 gm/mi	15 gm/mi	2.0 gm/mi
	0.41 gm/mi C	9 gm/mi C	1.5 gm/mi C
1978	0.41 gm/mi	3.4 gm/mi	0.40 gm/mi

Evaporative losses — 6.0 gms per test in 1970 in California, and 1971 nationwide, and 2.0 gms per test beginning in 1972 nationwide.

*Empirical mass equivalent for a 4000 lb. vehicle

C California requirement only

CALIFORNIA EMISSION CYCLE

This series of tests was widely used during 1966-72 for emission testing. It required the following seven modes to be run through seven complete cycles. The vehicle was operated on a chassis dynamometer to simulate road loads and samples of exhaust gases were collected and analyzed. Continuous strip-chart recordings were made of CO, HC and NOx concentrations in the exhaust.

There are seven cycles of seven modes. The modes consist of one idle, two accelerations, two constant speeds and two decelerations.

1. Idle
2. Accelerate at timed rate to 30 MPH
3. Cruise at 30 MPH
4. Decelerate to 15 MPH
5. Cruise at 15 MPH
6. Accelerate to 50 MPH
7. Decelerate with closed throttle to 20 MPH, then decelerate to stop and idle.

The first four cycles (of seven modes) are called "Cold Cycles" and are weighted 35%. The fifth cycle is disregarded. The sixth and seventh cycles called "Hot Cycles," are weighted 65%. The start is not included. The vehicle sits or "soaks" for at least 10 hours prior to the test at a temperature of 60-86°F.

Close-up of analyzers used to obtain final mass values of each gas, plus a continuous reading throughout the test. The analyzers signal a computer which gives a print-out (arrow). Computer gives the final test results and also a volume and mass concentration of each gas for each mode; i.e., idle, acceleration, etc. This mode read-out is a valuable analytical tool for the development engineer.

allows the crankcase to be vented through the air cleaner under low manifold vacuum conditions. Keeping this valve and restriction clean is essential to the correct operation of the engine because this is the only crankcase venting. The relationship of the PCV systems to the carburetor is a simple one—usually a tube connection or threaded connection in the carburetor base for attachment of the tube from the PCV valve.

The crankcase emissions were attacked first because approximately one third of all engine emissions come from this point. And, it was obvious from looking at the old road-draft vents that lots and lots of pollutants were being spewed into the atmosphere from this source. If you stand near an early car (pre-1965 in most of the U.S.), the smell of the escaping materials is quite obvious.

EXHAUST EMISSIONS

This area is more complex by far than the crankcase or evaporative ends of the problem. Exhaust emissions include unburned hydrocarbons, carbon monoxide and oxides of nitrogen.

In the 1966 model year, California applied standards for tailpipe emissions of CO and HC, together with test procedures and sampling methods. The California emission test was conducted by operating a vehicle on a chassis dynamometer. Actually, it attempts to simulate a 20-minute drive through downtown Los Angeles.

Actual exhaust emissions from the tailpipe are measured for concentrations in very sophisticated analyzers and recorded. These recordings are reduced to an average reading which gives more weight to certain operational modes.

The Federal Government adopted the California test procedures and standards in 1968 and required—as California had earlier—all manufacturers to certify and prove that their vehicles met the prescribed standards.

The Federal Government introduced a new test procedure beginning with the 1973 car model year. This procedure is called *Constant Volume Sampling* or CVS system.

The vehicle is "soaked" for 10 hours as in the California Test. The test is run on a chassis dynamometer loaded pro-

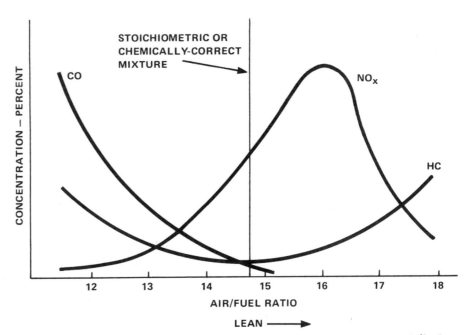

This plot shows the approximate relationship between CO, HC and NO$_x$ as air/fuel ratio changes. It shows the problem faced by the carburetor calibrator. As the mixture is leaned to produce low HC and CO emissions, the NO$_x$ emissions begin to increase. The best calibration is a compromise which tries to hold all three at an acceptable level.

portional to engine displacement and vehicle weight. A specific driving schedule consists of idle, acceleration, cruise and deceleration modes in a non-repetitive sequence. No two modes are alike. The test simulates 7.5 miles and runs for 23 minutes. The test fuel type is specified for uniformity. In this procedure the cold start is included in the test.

All of the exhaust gas is mixed with outside air and routed through a constant volume pump that mixes and measures the flow. A specified proportional part of the mixture is collected in a plastic bag. After the test is run, the contents of the bag are analyzed for concentrations of the three constituents (HC, CO, and NO$_x$). Because volume and concentrations are known, the actual mass of each gas can be calculated. The standards are written on a mass basis. Air in the test room is constantly monitored in a background bag. The mass of each constituent in the background bag is also calculated and subtracted from the values obtained in the sample bag.

The exhaust emission procedure was modified beginning with the 1975 car model year. The first 8-1/2 minutes or 3.6 miles of the test is rerun after a 10-minute engine-off period. This sample

is kept in a separate bag and called the *hot-transient portion*. An equal part of the first test is called the *cold-transient portion*. Its exhaust sample is also collected in a separate bag. These two sample bag values are averaged to allow for the fact that not all starts are cold starts.

The sample from the remainder of the first test is collected in a third bag and called the *hot-stabilized portion*. The average of the hot and cold transient portions are combined with this portion to yield the final exhaust-emission results.

Carbon monoxide forms whenever there is insufficient oxygen to complete the combustion process. Generally speaking, the richer the mixture, the higher the CO concentration. Even if the fuel/air mixture is chemically correct, CO cannot be reduced to zero because perfect mixing and cylinder-to-cylinder distribution is impossible to achieve.

Gasoline is composed of numerous and varied hydrogen and carbon compounds, hence the name *hydrocarbons*. Unburned hydrocarbons are just that—gasoline that did not get burned on its trip through the engine. There are several reasons why this can and does happen. Rich mixture is one. Fuel that does not get burned because of misfiring is another. So, either a lean or a rich F/A mixture can cause an increase in HC emissions. Other things affect HC concentration. Higher compression increases combustion-chamber surface-to-volume ratio and thereby increases HC emission. Spark advance and high vacuums under decelerations also affect HC.

Standards for oxides of nitrogen emissions were established by California for 1971. These became nationwide law in 1973.

Oxides of nitrogen (NO_x) are formed in the combustion chamber under high temperature and pressure conditions. Combustion pressures increase as the engine is loaded. The tendency to form NO_x is increased as the mixture is leaned because of the increased availability of free oxygen. Maximum NO_x production occurs under conditions of approximately 0.062 F/A (16:1 A/F).

EMISSION CONTROL APPROACHES

There are various approaches to the problem of making engines and vehicles meet the emission standards. The first approaches were engine modification and air injection.

As of 1972, another engine modification was being used: Exhaust gas recirculation or EGR. By 1975, most exhaust systems used either a thermal reactor or a catalytic converter. The thermal reactor appears in various forms, but is essentially a restriction in the exhaust system ahead of the muffler. Its purpose is to retain heat in the front part of the system, which encourages burning excess HC, and to reduce flow rate which gives more time to complete combustion of HC in the exhaust system. Thermal reactors are still another addition, used in combination with the already complex assortment of controls and gadgets to reduce emissions.

The catalytic converter, an addition to the exhaust system, assumes a major role in emission reduction as described in a later section. When a catalytic converter is used, some of the stringent controls applied to the engine can be eliminated, resulting in better drivability and fuel economy.

Engine Modifications—These include carburetor calibration and operational changes—along with distributor spark-advance settings, curves and operational changes.

Carburetors are set up with leaner mixtures in the part-throttle range. Mechanical limiters are placed on the idle-mixture adjustment screws to prevent excessively rich idle mixtures. Dashpots and throttle retarders, plus special idle setting solenoids are used. And, choking mixtures are eliminated very quickly after the engine has been started. Inlet air is heated to ensure good fuel vaporization and distribution.

Decelerations create very high manifold vacuums unless special controls are used. With a closed throttle, so much exhaust is sucked back into the intake manifold that the F/A mixture is diluted (leaned) to the point of borderline firing, causing missing and consequent high emission concentrations of unburned hydrocarbons.

Several controls can be used singly or in combinations:

1. Shut off the fuel flow so that there will be no unburned hydrocarbons—because there is no fuel entering the manifold. This method has the problem of creating a "bump" when the fuel is turned back on near normal idling manifold vacuum.

2. Supply a richer mixture to ensure burning, as is done by a deceleration valve and special deceleration fuel/air feed circuit on some Pintos.

3. Retard throttle closing to avoid high vacuum build up. Although this latter method is commonly used, it reduces the braking effect which would have been obtained from the engine during deceleration with a closed throttle.

Distributors are set up with more retard at idle and part-throttle and various switches, valves and other controls are used to provide advance or retard as required to meet emission requirements. Spark retard is an effective means of reducing HC because exhaust temperatures are increased so that the hydrocarbons are burned completely.

Additionally, basic engine modifications are being made, including reduction of compression ratio to reduce combustion pressures and temperatures, valve-timing variations and re-design of combustion chambers.

An additional engine modification, exhaust-gas recirculation, is discussed separately.

Sorting out which modification is accomplishing what is a bit difficult. However, let's over-simplify a bit and say that the retarded spark helps to reduce HC emissions by raising exhaust temperatures, hence ensuring complete burning. By the same token, retarded spark does not let the engine develop peak pressures which it is actually capable of, so NO_x is reduced simultaneously. Part of the combustion-chamber alterations which are being made include the elimination of flat quench surfaces as another way to slow burning and keep NO_x down. Also, by eliminating the cooler surfaces of quench areas, the HC emissions are dropped. Location of the rings on the piston has an effect on HC. If the rings are higher or the piston top is tapered toward the top ring—there's less volume in which unburned HC can "hide," thereby reducing HC still further.

Keeping the chamber hot also reduces deposits—which provide more places for unburned HC to "hide." These are all minimal things, but the game of reducing emissions is made up of a bunch of *small things*.

Keeping the compression ratio down reduces peak pressures—reducing NO_x.

Reducing emissions is a super balancing act—just like tightrope walking in many ways. The major complexity facing the engine designers is the internal-combustion engine's tendency to produce more oxides of nitrogen whenever hydrocarbons and carbon monoxide are being reduced—and vice versa.

Air Injection—This system accomplishes additional burning in the exhaust ports and manifolds. The exhaust manifold actually becomes an oxidizing reactor. A slightly richer mixture is used and extra air is added to oxidize (burn) unburned hydrocarbons and carbon monoxide. An engine-driven air pump and an air manifold deliver air to each exhaust port. Although this is a costly approach, the slightly richer mixture which can be used on engines which are thus equipped makes the car more drivable with less tendency toward surging in the midrange.

The first major use of air injection was on most 1966-67 California cars. GM calls the system AIR for *air-injection-reactor*, Ford terms theirs the *Thermactor* system. The air pump is supplied with filtered air from a built-in air cleaner/filter or from the air cleaner on the carburetor. The pump is protected from exhaust gases by a one-way check valve which only allows air to flow toward the exhaust system. A vacuum controlled diverter valve shuts off the air flow to the exhaust system during deceleration (overrun) so that backfiring will not occur. This diverter valve dumps all of the pump output through an air muffler.

Some engines had been equipped with AIR or Thermactors ever since the method was first introduced. Most of the cars equipped with these systems were those with high-performance engines. As of 1972, the air-injection systems began to be used more widely to help meet increasingly stiff requirements for reduced HC and CO emissions.

A more efficient method of oxidation is to redesign the exhaust manifold into more of a reactor by restricting flow and providing insulation to keep the heat in. These two items provide more time and higher temperatures so that HC and CO will be further oxidized. The added restriction however, causes excessive loss of power and the higher temperatures require more expensive materials. The Mazda rotary uses the reactor approach, but to date, U.S. manufacturers have avoided it.

Exhaust-Gas Recirculation (EGR)—This is another engine modification, but we have placed it here because it began to be used later than the other modifications. The first two production uses of EGR with built-in plumbing to accomplish it were on 1972 California versions of some Buick and Chrysler automobiles. Buick meters 1% to 10% of the exhaust back into the intake area below the carburetor to dilute the incoming fuel/air mixture, lowering peak combustion temperatures to reduce NO_x emissions. A passage in the intake manifold routes exhaust gas through a vacuum-controlled metering valve to holes in the manifold floor under the carburetor. The normally closed valve opens as the throttle is opened from idle thereby increasing the exhaust flow into the intake manifold.

Chrysler included a slightly different EGR system on their models. One or more orifices in the intake-manifold floor connect to the exhaust heat passage.

Three vacuum-controlled EGR valves. Chrysler valve on left, GM at top, Ford on right.

EGR valve with a tapered pintle giving an opening proportional to the vacuum signal applied to the diaphragm.

Typical EGR installation on a 1973 Chrysler. Valve transfers gas from exhaust gas cross-over to intake manifold.

EGR amplifier used on 1973 and later Chrysler. Carburetor venturi vacuum is the input signal. Through a complex system of valves and diaphragms the signal modifies manifold vacuum to provide a stronger signal to the EGR valve. Four hoses are for venturi-vacuum in, manifold-vacuum in, EGR signal out and a temperature-override signal that switches output off at lower engine temperatures. The latter function is not always used.

Bottom of 1975 Dodge Monaco has catalytic converter tucked up beside the frame upstream from the muffler. Shield protects converter from damage and helps deflect heat. Exhaust pipe is shielded top and bottom for the same reasons. Tap upstream of converter provides access to pre-converter exhaust gases for tuning and analysis.

These openings range from two 0.030-inch-diameter ones to a single one with 0.125-inch diameter.

Beginning with the 1973 car model year, virtually all vehicles had an EGR valve of one type or another. Many 1972's also had EGR valves. The valve controls exhaust gas routed either from the exhaust cross-over in the intake manifold or directly from the exhaust manifold and introduced into the intake manifold either directly or through a spacer beneath the carburetor. In all cases, care is taken by the engineers to distribute the exhaust gas equally to all cylinders. This becomes more and more important as EGR rates are increased.

These valves are of many types and configurations, but can be easily recognized because of the large operating diaphragm. Some are of the simple on-off type and others are tapered to increase exhaust gas flow with engine through-put. The trend is to use the latter type. There is also a trend toward valves which sense exhaust back pressure to make the recirculation rate more proportional to actual mass flow rate through the engine.

The carburetor usually supplies the signal for these valves either from a probe in the venturi or from a port in the throttle body. This port is not exposed to manifold vacuum at idle, much the same as a spark port, to prevent EGR at this point. The venturi pick-up gives a signal proportional to air flow through the carburetor. In some cases the carburetor signal is multiplied by routing it through a mechanical vacuum amplifier. EGR at heavy loads is unnecessary because NO_x does not form readily at richer power mixtures. This is handled automatically in some cases because the vacuum signal to the valve drops off with load.

The carburetor must also be considered because drivability is adversely influenced as EGR rate is increased. Simply richening the mixture is not the answer because this increases HC and CO emissions. (See diagram.) Efforts must be extended to improve vaporization, mixture preparation and cylinder-to-cylinder distribution.

Catalytic Converters—Catalytic converters were used for the first time on 1975 model year vehicles. Some of these also used an air pump. The catalyst is more effective when used in conjunction with an air pump, but cost is added to the system. This extra efficiency is not always needed to meet the 1975-76 interim standards, but will most certainly be required in the future. A catlyst and air-injection combination can have an efficiency as high as 90% under optimum conditions.

A catalyst is an element or compound that helps promote a chemical reaction but does not actively take part in it, that is; it is not itself changed chemically. These catalysts are of the oxidizing variety: They help oxidize the HC and CO into water (H_2O) and carbon dioxide (CO_2).

Most auto catalysts are the PGM (platinum group metal) type which includes platinum, palladium and ruthenium. These metals are mounted on beads or monolith honeycombed supports made of less expensive material such as alumina. The catalyst is installed in a can somewhere in the exhaust system, usually under the floor.

The carburetor supplies slightly richer part-throttle mixtures to assure a reaction and this improves drivability somewhat. The distributor is advanced more than non-catalytic packages which helps both drivability and fuel economy. In other words, engine modifications are lessened because HC and CO are handled by aftertreatment.

It is yet unclear as to just how often the catalyst will have to be replaced, but one thing is certain. *The catalyst must be used with unleaded fuel.* Lead in the fuel will "poison" the catalyst in just a few miles, rendering it ineffective. This could be a serious problem in remote areas and small towns where unleaded fuel may not always be available.

Evaporative Emissions—Here is the third major area for discussion. Evaporative emission includes hydrocarbon materials contained in fuel spilled or evaporated from fuel tanks and carburetors. While you might consider this to be an insignificant part of the emission picture, it is estimated evaporative emissions are equal to the emissions which would occur if crankcase ventilation were uncontrolled.

The gas tank and carburetor were traditionally vented to the atmosphere—until 1970 in California and 1971 nationwide.

The venting got rid of vapors, thereby aiding the hot-starting capabilities of the cars. There was no real concern about the emissions caused by spilled fuel during overfilling or caused by gasoline sloshing out of vents during sudden manuevers.

The most obvious part of this solution is the charcoal canister—because it is quite visible in the engine compartment. But, sophisticated measures are used in the fuel-tank-filling and tank construction areas, too. Thermal expansion is provided for by incorporating means to trap as much as three gallons of air during tank filling. Ford cars use a limited-fill method, whereas GM and Chrysler have another tank area or bell inside the tank. These tanks trap the air during filling, then allow it to escape to the top of the main tank where it is vented to a charcoal canister. The vent lines are very special as they incorporate vapor separation devices—small tanks located near and slightly above the main gas tank. Any liquid trapped therein drains back to the tank, but the vapor passes on to the canister—or, in the case of some Chrysler products, to the crankcase for storage.

In some instances, vapor-collection lines connect the carburetor to the charcoal canister. The canister, incidentally, is about the size of a small cookie jar and located wherever room can be found in the engine compartment. Fuel vapors are adsorbed onto the surfaces of the charcoal granules. When the car is restarted, the vapors are sucked into the carburetor or intake manifold to be burned in the engine. The method of purging the canister varies widely with make and model. In some instances a fixed restriction allows constant purging whenever there is manifold vacuum. In others, a staged valve provides purging only at speeds above idle. In 1972, GM incorporated a thermal delay valve so that the canister is not purged until the engine has reached operating temperature. Purging at idle or with a cold engine creates other problems, such as rough running and increased emissions due to the additional vapor being added to the intake manifold.

In 1970-71, total evaporative emissions allowed from any car were six grams per hot soak. In 1972, the allowance was reduced to two grams per hot soak. It should be noted here that this reduction

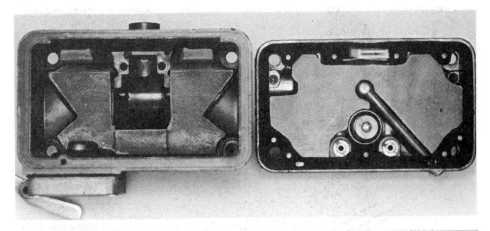

The large bowls of the Holley, although an advantage for performance and for hot-starting capability, have to be reduced in volume with these bowl and metering-block stuffers to allow meeting emission standards on some vehicles. Both California and Federal specifications spell out total emission specifications for whole-vehicle evaporative loss, that is, for the combination of all parts which can vent fumes into the atmosphere: Air cleaner, fuel tank, etc. The stuffers serve no other function except volume reduction and they should not be used for racing applications.

Heated inlet air improves vaporization and thereby aids distribution. It allows using leaner fuel/air mixtures to reduce emissions. Thermostatic valve modulates temperature of air entering carburetor by opening flapper to admit ambient air when underhood temperature exceeds 85°F. The heated-air supply is closed off by the flapper when the underhood temperature exceeds 125°F. Chrysler illustration.

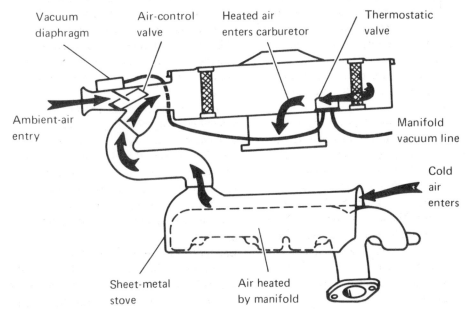

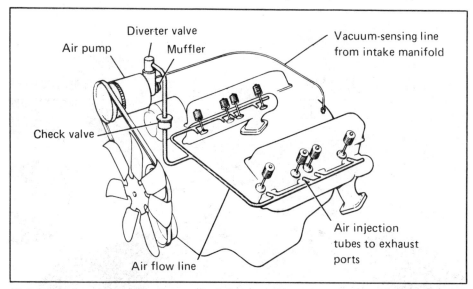

Air-injection systems are similar on all of the cars which use them. This is the Chrysler version used on 1972 California cars. Pump adds a controlled amount of air to the exhaust gases in the exhaust manifolds to ensure complete burning. Chrysler drawing.

in allowable evaporative emissions created the need for reducing the bowl capacity of some Holley carburetors with bowl stuffers.

Altitude—Beginning with the 1977 car model year, all vehicles sold at altitudes above 2,500 feet must meet emission standards at the altitude at which the vehicle is sold. In most cases the carburetor will have a different calibration for altitude.

EMISSION SETTINGS

Currently all manufacturers are required to install a specification label in the engine compartment. It details the correct carburetor and distributor adjustments needed to maintain legal emission levels for that vehicle. No matter how effective or sophisticated the control system, a weak spark, fouled plug, bad plug wire, cracked distributor cap—or any of many other things—can wreck the system's efficiency. It only takes one of these things going wrong to cause the car to become a pollutant emitter of a worse nature than a car with no controls at all.

A single bad spark plug is a good example. In SAE Paper 710069, "Exhaust-Emission Control for Used Cars," the authors pointed out that a single fouled plug increased the HC emission level six times: From 605 ppm to 3,609 ppm.

WHAT'S IT COSTING?

The whole idea of emission controls is to clean up the environment—namely the air we breathe . . . and live in. But, as is always the case when you get down to the facts—there is "no free lunch." Anything you get—in this case cleaner air—costs *something*. That something includes reduced drivability, increased gasoline consumption, increased tune-up costs—and a car that weighs more.

To these obvious costs we must add increased complexity of the entire vehicle with associated cost increases in both engineering and manufacturing.

Fuel-Supply System

Ronnie Sox checking the throttle-return spring on his Chrysler-hemi-powered drag car. Note use of two vacuum-secondary Holleys on factory plenum-ram manifold. Arrows indicate two fuel filters—one in the line to each carburetor.

AVOID THESE FUEL-SYSTEM PROBLEMS

Racers often get into leaning-out problems which are no fault of the carburetor or its jet combination. Such situations can usually be identified without too much effort because increasing main-jet size will not help the problem and will probably worsen it at low and mid-range RPM. The problem is insufficient fuel pressure.

Inadequate fuel supply at the carburetor can be caused by:

1. Pump with insufficient capacity

2. Restrictive fittings or fuel lines too small, inlet seats too small

3. Too many bends, or a crimped line

4. Tank outlet or filter screen restricted

5. Plugged fuel filter/s

6. Use of filter between tank and pump instead of between pump and carburetor

7. Dirt anywhere in the system

8. Line too near a heat source causing boiling in the line.

FUEL PRESSURE

Fuel pressure should be measured with a fuel-pressure gage which reads from 0 to 10 psi. Checking the fuel pressure is just one more of those smart things to do to avoid wasting a lot of time and money applying corrective measures to areas which are not contributing to the problem. Measuring pressure just ahead of the carburetor ensures that you will know what fuel pressure is available at that point.

The fuel-pressure measurement that is most important is the one which is made just as the engine is at its RPM peak—for the course—where it is using the most fuel. The pressure reading at this point should be at least 3.5 psi, but any pressure from 3.5 to 8 psi is OK. 6 psi is considered a normal fuel pressure by Holley engineers.

Any pressure lower than 3.5 psi is not acceptable because this indicates that the pump or fuel-supply system is not keeping up with the engine's fuel requirements. Pressure above 8 psi should be approached with great caution and accompanied with appropriate fuel-level changes. Pressures over 6 psi overcome the closing force applied to the inlet needle valve and create an unnecessarily high fuel level in the fuel bowl. Each added psi raises fuel level approximately 0.020 to 0.030 inch over the nominal leve established at 6 psi.

The fuel-pressure gage must be mounted so that the driver can see it without having to take his eyes off of the course. Drag

racers mount their fuel pressure gages on the cowl outside of the windshield because common sense and drag rules do not permit fuel lines or containers in the driver's compartment.

In case it's not immediately obvious why the fuel system should be routed outside of the driver's compartment, just think about what happens when a clutch, flywheel or transmission (standard or automatic) explodes and cuts any part of the fuel-system plumbing. Sparks are sure to occur at some point during such an incident. Similarly, if a car is crashed, the danger of fire is always present and you want to be sure that you provide as much safety margin as possible for yourself or your driver. We lost a very good friend this way in 1960—Wayne Ericksen.

If you insist on a fuel-pressure gage in the cockpit, then use a remote pressure sender (electrical) at the point where you would take off pressure for a mechanical pressure gage—at the intersection of the Y

Holley canister-type fuel filters with tube and hose-type connections are neat additions to any fuel system. Replaceable elements are available (center).

which feeds both float bowls (on carburetors with two inlets) or at the connection to the carburetor.

FUEL LINES

Fuel lines must be routed away from heat and firmly secured so that they will not vibrate and fatigue.

Fittings must be non-restrictive. Some fittings have built-in restrictions. Check every fitting that you use in your fuel-supply system. Make sure that the passage is the same size all the way through the fitting, you may be able to open up the passage with a drill.

Avoid 90° (right angle) fittings wherever possible as these give the most fluid friction and therefore are more restrictive than straight-through or 45° fittings. A right-angle fitting has the same restrictive qualities as a piece of tubing which is several feet long.

If a mechanical fuel pump is used, install a 1/2-inch ID line from the tank to the pump. Make sure that the fitting which connects to the fuel tank is as close to the fuel line ID as is consistent with safety. Fuel-line attachment to the fuel tank is one area which is commonly overlooked by mechanics who are just getting started in competition. They will install a large line, but try to feed fuel through a tiny 1/4-inch or smaller diameter opening, which does not work. Here is one place where steel, copper or aluminum tubing is a better choice than fuel hose because tubing fittings typically allow a larger thru-passage than you can obtain with hoses. This is because the hose fitting has to have enough wall thickness to withstand the clamping pressure applied by a hose clamp. Fittings for tubing can have openings which are very close to the actual ID of the tubing itself, thus giving less restriction.

Remember that the important point is to keep the fuel supply to the pump unrestricted. The pump outlet (for an engine-mounted mechanical pump) or regulator outlet connection to the carburetor/s can be through 5/16- or 3/8-inch tubing.

FUEL FILTERS

A fuel filter should be used between the fuel pump and the carburetor. Never use any kind of filter—other than a simple screen—on the suction side of a fuel pump. This is true regardless of the fuel-pump type which is used. The usual screen at the tank outlet will usually be o.k. if it is clean.

The filter in the line to the carburetor should be as non-restrictive as possible. Paper-element filters such as those sold by Holley are excellent for the purpose. If you are concerned about the pressure drop through the filter, "Y" the fuel line to run through two filters so each one supplies a carburetor, or use two filters in parallel in the line. Mount the filter canister to allow easy replacement of the element.

Sintered-bronze (Morraine) filters are found in the inlets of many two- and four-barrel carburetors. These are OK for street use if there is little dirt in the gasoline and if the tank itself is clean. If the fuel supplied to the pump is dirty—as may be the case in a dusty area—

Below: Holley high-performance fuel pump for small-block Chevrolets, P-4767A. Delivery of 80 gallons per hour at 7000 RPM is typical for all these mechanical fuel pumps. Bottom photo is a Holley pump for small-block Fords, P-4766A disassembled to reveal construction details. All of these pumps feature large inlet/outlet ports (allowing use of hose to 3/8-inch ID) in rotatable housings which can be turned through 360° to allow easiest plumbing routing. Repair kits are offered for all Holley pumps.

remove the sintered-bronze filters from the inlets and install an inline filter between the pump and the carburetor.

Pay attention to the fuel filter and replace the paper element or throwaway inline filters when the fuel pressure drops. This indicates that the filter has done its work and it is time for a new one.

For competition, the sintered-bronze filters should be taken out of the inlets and an inline filter installed in the fuel-supply system.

FUEL PUMPS

There are two common types of fuel pumps: Mechanical and electrical. Although the carburetor never "knows" which kind supplies its fuel, let's consider the two types briefly.

Engine-mounted mechanical pumps are diaphragm types driven by camshaft/crankshaft eccentrics operating a rod or lever. Advantages include low first cost, simple plumbing and mounting, low noise level—and familiarity of the general public with the type because millions have been used over the years. Disadvantage is transferral of engine heat into the fuel, especially when the engine is stopped. The mechanical pump sucks fuel through a long line from the tank, further promoting the fuel's tendency to flash into vapor, especially on warm days.

Electrical pumps are not widely used as original equipment except for the in-tank pump on the Vega. The electrical pump has a higher first cost, is noiser than a mechanical pump (unless special mounting procedures are used) and requires connection to the car's electrical system. It must be plumbed to the tank outlet and to the fuel line.

When installing an electrical pump, it is typically mounted at the rear—near the tank—as a "pusher" device. Fuel pushed forward to the carburetor has less tendency to flash into vapor as it moves through the line. For competition or hot-weather the rear-mounted electrical pump provides "vapor lock insurance." There are several types of electrical fuel pumps: Fixed-vane, sliding-vane, and solenoid operated. The Holley GPH-110 uses sliding vanes.

When installing an electrical pump, eliminate the mechanical pump if at all possible. It heats the fuel and limits pressure because it acts as a restrictor in the line from the tank.

High-pressure version of Holley pump pushes fuel from tank at high pressure to regulator mounted near carburetor to establish desired fuel pressure delivered to carburetors. Essential component for pump installation is pressure switch 89R-641A in center. It senses engine oil-pressure to shutoff fuel pump when engine is stopped, whether ignition is turned off or not.

COOL CANS

The need for cool cans for competition has been discussed on page 115. One of these devices is pictured on page 120 of the performance chapter.

HOLLEY GPH 110 PUMP

This electric fuel pump is a positive-displacement type. Pumping action is accomplished by four sliding vanes spaced in a rotor at 90° intervals. Vanes are thrown out against the rotor chamber walls by centrifugal force. The rotor is offset in its pumping chamber so that the inlet side is larger than the outlet side. A pumping segment created by two vanes picks up fuel as the segment passes the inlet port. Vanes move the fuel from the "larger" side of the chamber to the "smaller side. Dead-head pressure (pumping against pressure built up in a closed-end pipe) is regulated by an internal spring-loaded bypass valve which routes fuel from the outlet to the inlet side.

Power for the pump is supplied by a 12-volt permanent-magnet-type DC motor which runs dry—that is, no fuel is in the motor portion of the pump. A seal on the motor shaft keeps fuel out of the motor. Bronze bearings are used at each end of the motor shaft.

Both low-pressure and high-pressure (high performance) versions of the pump are offered. Although most of the other components are the same, the low-pressure pump has a different bypass valve and a motor with a lower rating and a different winding. This pump is dead-headed at 7 psi and is designed for street and highway use without a fuel-pressure regulator. It is not designed for competition.

The high-pressure pump is supplied with a remote fuel-pressure regulator which should always be mounted near the carburetor/s. A spring operates against the regulator diaphragm to open the internal valve. Line pressure against the same diaphragm closes the regulator valve to maintain a preset pressure. Regulated out-put (downstream) pressure can be externally adjusted from 3 psi to about 10 psi, dead head by changing the spring-preload adjustment screw. A locknut maintains the setting. All pump and regulator inlets and outlets are tapped for 3/8 NPT.

When installing the pump, include a safety switch in the circuit so that the pump will not work unless there is oil pressure. Holley's 89R-641A switch can be mounted on a 1/8-inch pipe tee (the stock pressure switch for the idiot light

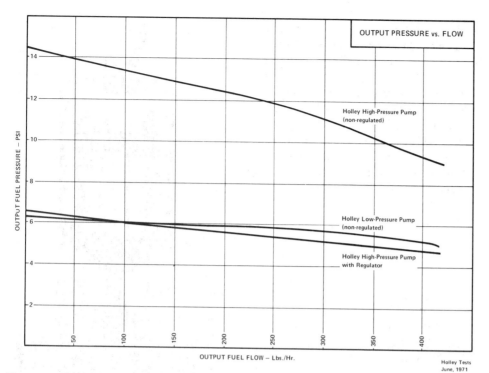

OUTPUT PRESSURE vs. FLOW

Holley High-Pressure Pump
(non-regulated)

Holley Low-Pressure Pump
(non-regulated)

Holley High-Pressure Pump
with Regulator

OUTPUT FUEL FLOW – Lbs./Hr.

Holley Tests
June, 1971

Top curve shows dead head pressure of 14.5 psi. Pump can supply 420 lbs/hr (75 gallons per hour) at 9 psi regulated output pressure. When regulator is set for 6.5 psi output, 420 lbs/hr is maintained while the pressure drops only to 4.75 psi.

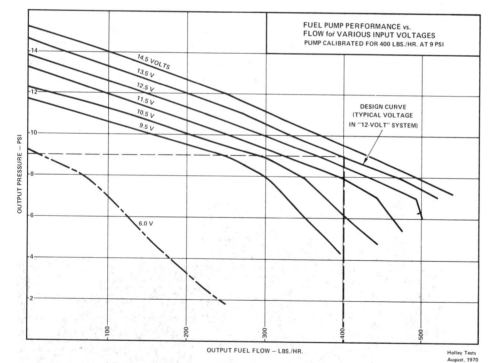

FUEL PUMP PERFORMANCE vs.
FLOW for VARIOUS INPUT VOLTAGES
PUMP CALIBRATED FOR 400 LBS./HR. AT 9 PSI

14.5 VOLTS
13.5 V
12.5 V
11.5 V
10.5 V
9.5 V

6.0 V

DESIGN CURVE
(TYPICAL VOLTAGE
IN "12-VOLT" SYSTEM)

OUTPUT FUEL FLOW – LBS./HR.

Holley Tests
August, 1970

Note how pump performance drops off as voltage drops, emphasizing the importance of keeping voltage up. The pumps are designed to operate at 13.5 volts — the regulated voltage of nominal 12-volt systems. At 13.5 volts the pump draws approximately 4 amperes.

goes on the other side) so that the pump will shut off if the engine stalls with the ignition on. Using the switch to turn the engine with the starter energizes the pump from the starter solenoid circuit. Once the engine is running, the switch provides voltage to the pump so long as there is oil pressure to keep the switch turned on. A schematic showing the wiring for the switch is included here for reference.

Mount the pump in as cool an area as possible—away from any exhaust-system components. Any fuel pump mounted to the chassis will transmit some noise into the car's body structure. This is no problem on a race car with open exhaust, but it can be a source of annoyance on a dual-purpose car which is driven on the street. The optimum way to reduce the transmission of pump noise into the body is to mount the pump on rubber-insulated studs which are mounted in turn on a chassis member or on a stiffened section of the body. Never mount the pump directly onto a large flat sheet-metal surface because the result will be amplification of the noise—just the opposite of what you are trying to achieve.

The pump should always be located near the fuel tank (usually at the rear of the car—unless the engine happens to be back there instead) as low as possible so that suction height will be kept to a minimum.

Creating a vacuum (low absolute pressure) on the end of a fuel line, especially a long one, tends to allow the fuel to flash into vapor—which is very hard to pump. This is why professional racers replace the engine-mounted mechanical pump with an electric pump near the fuel tank—it's positive insurance against vapor lock.

When fuel is pumped forward at high pressure, there is another advantage which may not be quite so obvious: This tends to offset pressure losses caused by line loss (friction) and by acceleration (g forces). Pumping the fuel forward at high pressure and then reducing its pressure prior to sending it on to the carburetors ensures that there will always be adequate pressure available at the carburetors.

Keep the pump-inlet line short and large. If a filter is used between the fuel tank and the pump inlet it must be a screen-type filter so that there will be no pressure drop at this point—which could cause fuel vaporization and consequent pump cavitation. It is actually preferable to position the pump so that its inlet is slightly below and behind the tank so that the fuel level will create a "head" or positive pressure at the inlet and so that acceleration forces will tend to keep fuel always at the inlet.

Stock-size (5/16 or 3/8-inch) fuel lines can be used on a street machine, but the pump performance will be compromised by such an installation. Small lines, sharp bends or kinks or right-angle fittings will always cause a pressure drop.

Because only one of these pumps is needed for adequate fuel delivery—even on a Pro-Stock race car—part of the money that would have been spent for a second electric pump can be applied to the purchase of large fuel lines.

Keep the lines away from exhaust-system components to avoid excessive heat. Make sure that there is no part of the body that will deflect the exhaust from open headers back into the area where the line is mounted. If a line passes near the exhaust system and there is no other place to route it, insulate the fuel line very thoroughly. The line should be clamped against the chassis or body structure with rubber-lined clamps such as those which are used in aircraft repair shops.

Locate the pressure regulator as close to the carburetor as possible. Lines between the regulator and carburetor/s can be 3/8-inch ID. There are two outlets on the Holley regulator. If the carburetor has fuel bowls with individual inlets, connect each bowl to an outlet. Where two carburetors are used, connect each carburetor to an outlet.

Pressure at the carburetor must be set with the engine idling so that there will be some flow to allow the regulator to function. Use a fuel-pressure gage at the carburetor and adjust the regulator to provide the desired pressure. 6 to 7 psi output is the pressure set at the factory when the regulator is tested.

When running the car at high ambient temperatures, a cool can should be used just ahead of the regulator (upstream) on the high-pressure side so that the fuel will not tend to flash into vapor when it is changed to a lower pressure by the regulator.

Some racers install a separate bypass from the pump outlet to the tank. Then the pump does not continually pump the same fuel in a loop from the outlet back to the inlet. The external bypass allows fresh fuel to appear at the inlet, even in low-demand situations such as idling. The bypass should be through a 1/16-inch (or 0.060) restriction. Even with this bypass arrangement, the pump capacity of the high-performance version is 360 lbs./hr. of fuel at 9 psi. Don't use any restrictor larger than this!

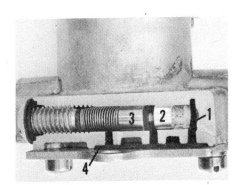

Passages in the pump casting provide regulation of output-pressure. Cutaway of pressure-relief valving shows how this all works. Fuel at higher than desired pressure from the outlet side 1 enters a hollow seat 2 to act against the spring-loaded piston 3. Pressure pushes the piston off the seat so high pressure fuel is routed back to the inlet side of pump 4, thereby regulating pressure. Note how rotor 5 is offset in pumping chamber 6 formed by vanes 7 in wear sleeve 8. Cutaway side view of pump shows rotor 5 with sliding vanes 7, wear sleeve 8, wear plates 9 on top and bottom of pumping chamber, seal 10 and bronze bushing 11 for armature shaft.

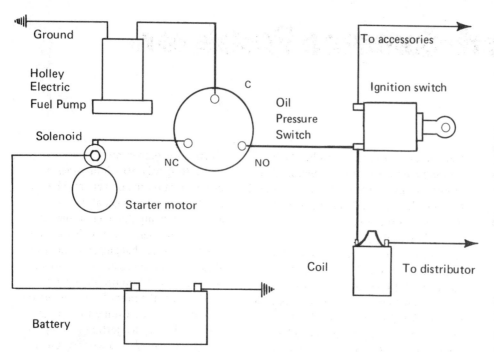

Holley pump installation in Wally Booth's Pro Stock Camaro shows a by-pass line installed in the location normally occupied by the screw for the pressure-relief valve. Although hose is large, a restrictor is used as described in the text. Arrow points to the by-pass line.

Typical wiring diagram for an electric fuel pump. Holley's pressure switch is used to ensure that pump will not operate unless the engine is running — or the ignition switch is being used to energize the starter solenoid. The starter circuit from the ignition switch is not shown in this drawing.

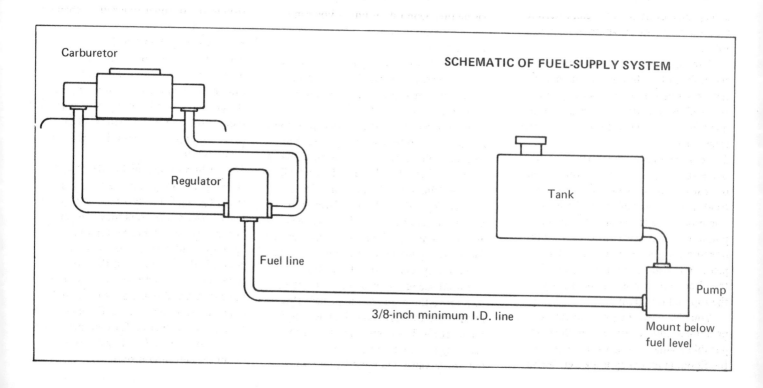

Racers' 10 Most-Common Problems

Holley has been a major sponsor with representation at races across the country for many years. This includes circle track and drag racing. Our tech reps talk with thousands of racers at our tents where we physically work on carburetors—and in our Holley Clinics. From these contacts we have compiled the ten most common problems. These are slanted toward the drag racer, but some are common to road and track racing.

1. **Dirt or rust in the system**—Remove the fuel bowl and there it is. Good old ferrous oxide (rust). Looks just like Georgia mud. The cause is a corroded fuel tank. The tank might be one that sat for a while in a high-humidity environment or one purchased second or third hand from a "buddy" who may not have known it was rusty or dirty. Rust is the most common foreign material to invade the fuel system but sometimes plain old dirt finds its way in through poor-fuel-handling practice. Whatever the contaminant, flooding usually results because something gets between the inlet needle and seat. Electric fuel pumps are also susceptible. Contaminants can interfere with operation of the internal regulator valve.

A temporary fix is a new unrestrictive in-line filter, but eventually a dirty tank must be flushed and a rusty one replaced. By the way, always use an in-line filter as a safeguard even with a new clean tank.

2. **Fuel-pressure**—The most common one is running out of fuel at the top end or near maximum wide-open-throttle RPM, where fuel demand is the highest. The problem is either an inadequate pump, restrictive system or perhaps the pressure was set too low in the first place. You need a pressure gage to find this. Remember, pressure below 3 psi at the top end is marginal.

You also need the pressure gage to set pressure at idle for systems with an adjustable regulator. If pressure is too high the flooding tendency is greater. We've

seen racers try to run the Holley high-pressure pump without a regulator. The result is 15 psi to the inlet and guaranteed flooding. See the fuel-supply chapter for more information.

3. **Incorrect float level**—Correct procedure for setting the externally adjustable floats is covered in detail in the Model 4150 Repair & Adjustment section. Most racers use carburetors with these bowls. Remember fuel level should be reset after changing the fuel-inlet valve or fuel pressure. The fuel level must be checked on a new carburetor because shipping shocks can alter the float setting.

4. **Accelerator pump incorrectly adjusted**—The correct method is shown in the Model 4150 Repair & Adjustment section. Remember the pump-operating lever should contact the pump lever at idle (or at staging RPM throttle opening) and the pump lever should have a little travel left at wide-open throttle. When you've done this you've got a full stroke. We've seen bent pump-operating levers. Someone probably wrote an article advising this. Don't do it unless you can't get a proper adjustment with the screw—which is rare.

5. **Use of non-Holley parts**—Most common are gaskets, jets and inlet valves. Some are pretty good, but most are not up to our standards. Some gaskets are made of less-expensive material—probably not tested with typical racing fuels. Not all non-Holley jets are flow-checked. Many so-called "high-flow" inlet valves actually flow less fuel than Holley items. Stick with Holley parts in Holley Carburetors.

6. **Removal of power valves**—We see an alarming amount of this. The only reason for removing the power valve is where the induction system is so restrictive that manifold vacuum at wide-open throttle gets above the rating of the power valve. In this case the valve will close, leaning the mixture. The correct solution is to use a higher rated valve. If you remove the power valve you must jet up 6 to 8

jet sizes to compensate

7. **Jetting way off**—Sometimes this occurs because the power valve has been removed and things get all twisted around. Reading the plugs as described in the performance section helps guide you in your jet changing, and can show if stagger jetting is necessary to correct a fuel/air distribution problem. Once you've "lost the rabbit" for one reason or another the best thing to do is go back to the original jetting with the correct power valve/s and work in small increments from there.

8. **Screw in secondary-throttle lever on diaphragm-operated carburetors**—This is done to convert a diaphragm-operated secondary to a mechanical one. The problem is, there is seldom enough pump shot to get you through without a bog on a wide-open-throttle "punch." Worse yet, in some cases the throttle can jam open and ruin your whole day. Don't do this little trick. Buy a double-pumper if you want mechanical secondaries—or play with the secondary diaphragm spring to get the opening characteristics you want. Use a stop watch or timer to evaluate changes. "Seat of the pants" feel can be deceiving.

9. **Dirt-clogged air bleeds**—When this happens the metering of the idle and/or main systems is altered. Sometimes these can be freed with a little spray-on solvent. If they are clogged badly you should remove the bowls and metering blocks, spray with a solvent, and then blow the passages out.

10. **Carburetor just plain "tired"**—It's amazing how long some carburetors stay around going from car-to-car and hand-to-hand. After a few years they can get into pretty bad shape. You have to make up your mind whether to replace or repair. If the carburetor is the size and type that is just right for your car and the castings and shafts are sound, you might want to repair it. Read the Repair & Adjustment section for your particular carburetor. Remember to get a Holley repair kit before you start.

Carburetor Adjustment & Repair

ANALYZING THE PROBLEM

There are three reasons why you would take wrench and screwdriver in hand and lay your carburetor open from choke plate to idle screw.

1. You are curious.
2. You're a racer or super enthusiast.
3. You've got a driving problem with your vehicle and want to fix it.

If you are a number one or two, skip to the appropriate section on disassembly and repair. If you belong to group three—this section was written *just* for you.

How can you be sure the problem lies solely within the carburetor, or involves it at all? An old carburetor man I know claims he has fixed a lot of "carburetor problems" by changing spark plugs and distributor points. There is a lot of truth in that statement.

When you are analyzing engine malfunctions you have to look at the *total* system. In the next few pages I do just that. Common problems associated with carburetors are presented along with some probable solutions. Look under the problem list for the malady that is occurring. Numbers there refer to possible answers in the solution list. As noted earlier, I also list some non-carburetor causes of a problem for a simple reason. Disassembly and repair of a carburetor can be quite time consuming. It sure is disappointing to go through the whole thing only to discover *later* that the problem is in some other component or system. My objective is to get you to the solution as quickly and simply as possible.

PROBLEM LIST

A. Hard starting—Cold engine
   1, 2, 3, 4, 5, 6, 7, 8, 9, 10, 11.
B. Hard starting—Warm engine
   1, 10, 12, 13, 14, 15.
C. Rough idle and stalls
   4, 7, 8, 16, 17, 18, 19, 20, 21, 22, 33, 53.
D. Deterioration of fuel economy
   1, 4, 7, 18, 20, 23, 24, 25, 26, 30, 49, 52, 53.
E. Sag or hesitation on light accelerations
   8, 10, 14, 18, 24, 25, 27, 28, 30, 32.
F. Sag or hesitation on hard accelerations
   23, 27, 29, 30.
G. Surge under cruising and light loads
   8, 10, 14, 18, 24, 25, 26, 30, 31, 32.
H. Surge at high speeds and heavy loads
   23, 25, 26, 31, 34, 35.
I. Misfire or backfire
   1, 4, 21, 27, 31.
J. Loss of power and top speed
   1, 14, 18, 20, 23, 31, 34, 35, 36.
K. Fast idling or inconsistent idle return
   8, 19, 37, 38, 39, 40, 41, 42, 43, 44.
L. Vapor lock—Loss of power, stall or surge on acceleration after short engine-off period in high temperatures
   10, 34, 35.
M. Inoperative secondary system (Diaphragm-Operated)
   45, 46, 47, 48.
N. Engine bucking or emitting black exhaust smoke
   1, 23, 49, 50, 51, 52.
O. Poor cold driveaway
   1, 2, 3, 29, 30, 51.

SOLUTION LIST

1. Binding or sticky choke plate or linkage. This can sometimes be repaired without removing the carburetor. If you have a remote or divorced choke, disconnect the choke rod from the carburetor to isolate the problem. *Never use lubricating or household oil to free the choke.* Oil gathers dust and grime and stiffens at low temperatures, creating a worse problem than you started with. Commercial solvents for carburetors work well. Squirting on penetrating oil works just great.

Sometimes vibration or a few backfires will cause the choke plate to shift on the choke shaft. Repairing this requires loosening the choke plate screws and realigning the plate in the air horn. On Holley chokes using screws with lock washers this is an easy task. Others have staked screws to prevent loosening and loosening these staked screws requires considerable care to avoid twisting off the screwheads. Use a small file to remove the upset or staked material from the screw and proceed with caution. You must restake the screws when you re-install them. The opposite side of the shaft must be firmly supported so you won't bend the shaft in the process. Disassemble the carburetor for staking the screws. With a divorced or remote choke the linkage or bi-metal in the choke pocket in the intake manifold can rub up against the side, causing binding. Move the choke rod up and down to detect this malady.

2. Incorrect choke adjustment. There are two basic adjustments—the bi-metal index and qualifying or pull. More information on your particular carburetor is in Section 3.

158

3. **Overstressed bi-metal.** If the choke unit is not exerting enough force to close the choke plate when the engine is cold and the ambient temperature is below 60°F (15.5°C), this could be your problem. A new bi-metal is required. Because replacement bi-metals are not always readily available, try resetting the choke in the RICH direction. Most automatic-choke housings have an arrow indicating RICH and LEAN directions.

4. **Fouled or old spark plugs.** You can easily spot this with an engine analyzer but not many of us have this luxury. Pull a plug or two and take a look. If you see a build-up of deposits or burned electrodes—change the *whole* set. Spark plugs are relatively inexpensive and the benefits of good firing are numerous, especially for economy. If plugs and wires are over a year old, replace the whole set regardless of how good they look.

5. **Wrong distributor point gap or dwell.** Detect this with a dwell meter or by physical measurement. With the point rubbing block set on one of the cam lobes, the points of most non-electronic distributors should be open about 0.017 of an inch. Check with a feeler gage or a No.77 drill. Check the points for pitting or build-up. Point changes can be accomplished on the engine but it is easier and less risky if the distributor is removed. If you remove the distributor note the position of the rotor and housing and have a timing light handy to re-time the distributor correctly when you re-install it. Also, you have to make sure to get the distributor drive mechanism fully engaged with the oil pump if it drives the pump.

6. **Water condensed in the distributor cap.** This can happen when there are warm humid days and cool nights. Remove the distributor cap and wipe with a clean dry cloth. Check the cap for cracks or carbon tracking while you have it off.

7. **Poor compression.** This could result from worn piston rings or leaky intake or exhaust valves. A compression or leak-down test will give a quick answer. Consult a service manual for desired compression values. If one or more cylinders has more than 20% lower compression than the others, you've got problems. A leak-down tester is better than an ordinary compression tester because it can help you find the leak.

8. **Intake leaks.** Look very closely at all of the rubber vacuum hoses. There are more and more of them these days. Make sure they are not split, leaking or disconnected. These hoses become hardened with age. Cracks are often hidden in bends or are on the underside, so do a careful job of looking. Be sure to check the hoses to the automatic transmission. Manifold-to-head-gasket leaks can be detected by spraying a little fuel all around the intersection while the engine is idling. If the engine speeds up or smooths out, you've found the culprit.

9. **Low battery voltage.** Slow cranking speed is the giveaway here.

10. **Fuel volatility.** Oil companies change the volatility or vapor pressure of their fuels so they are highest in mid-winter and lowest in mid-summer. This allows quick cold starts with little cranking in January and minimizes the long cranking to get started when the engine is hot in July. Volatility is also varied geographically with higher values in the North and lower ones in the South. An unusually warm day in the Winter always brings a rash of hot-starting problems. Avoid filling up at low volume stations—especially in the Spring and Fall.

11. **Water in the gasoline.** This is not as great a problem as it once was, due to the gasoline additives now being used. Nevertheless, always fill your tank completely. This minimizes the opportunity for water vapor to condense in the unused volume. This also reduces the likelihood of rust formation in your fuel tank. For the same reasons avoid purchasing fuel at low-volume stations. It is a good idea to add a can of water-absorbing fuel-tank cleaner like Dry Gas once in a while.

12. **Incorrect carburetor fuel level.** An excessively high fuel level causes rich die-outs on sudden stops due to fuel spillage. There is also a greater tendency toward fuel percolation (boiling over).

A low fuel level can cause sags on accelerations because it delays the main system start-up. Fuel starvation (leanness) on turns and maneuvers can also result from a low fuel level.

Some Holley carburetors have externally adjustable fuel levels and removable sight plugs. With the engine idling the lock screw should be loosened and the adjustment nut turned until the fuel level is just

at the bottom of the sight-plug hole. Other carburetors require disassembly for float adjustment. This is treated in more detail for specific carburetors in later sections. The Holley Spec Manual gives fuel-level adjustments for most popular Holley carburetors.

13. **Leaky fuel-inlet valve.** Is the valve worn or is foreign material lodged between the inlet and seat? New inlet valves and seats are included in most repair kits. Fuel levels are noted in the kit instructions. Always check the fuel level when changing fuel-inlet valves or floats.

14. **Sticking exhaust-manifold heat-control valve.** Most vehicles built prior to 1969 had heat-control valves. This valve diverts some of the exhaust gas through a passage in the intake manifold to create a "hot spot." If the valve is stuck in the closed position the hot spot will run at excessively high temperatures, causing vapor formation and boiling in the carburetor fuel bowl and hard starting and rough idling. If the valve is stuck in the open position the hot spot will not reach high enough temperatures, causing low and mid-range drivability problems.

The valve can be freed by using commercial solvents designed for this purpose. Penetrating oil can also do the job. WD-40 is especially good here. If the valve has been stuck for some time, it may take several applications, a little time, and some friendly persuasion with a hammer to free it.

15. **Mechanical bowl-vent adjustment.** This bowl vent is on pre-1968 carburetors. It should be slightly open (approximately 1/16") at idle and engine off and closed at all other conditions.

16. **Carburetor icing.** This occurs with high relative humidity and temperatures in the 30°-50°F (minus 1° to 10°C) range. While the engine is warming up, ice actually forms between the throttle plate and bore. Check the carburetor-air preheat system on 1969 and later vehicles. On earlier models check the exhaust-manifold heat-control valve (see 14). A slightly higher idle speed can help minimize this problem.

17. **Idle-mixture adjustment.** Specific carburetor sections show the location of the idle-mixture screw or screws for your particular carburetor. Turn the idle-mixture screw in until the engine speed drops

slightly, then turn the screw back out about 1/8 of a turn. This is referred to as the lean-best-idle setting. For 2- and 4-barrel carburetors, do this operation on one side and repeat it on the opposite side. Then go back to the original side and repeat the process one more time just to be safe. On 1968 and later vehicles, the idle-adjustment range is limited for exhaust-emission purposes.

18. Clogged air bleeds or passages. This usually requires a carburetor teardown just to determine if it really is the problem. Try swabbing and squirting the exposed bleeds with a little Gum-Out or other carburetor solvent or isopropyl alcohol. This may free things up and save a lot of work.

19. Idle-speed adjustment. Specific carburetor sections show the location of the idle-speed adjustment screw. 1970 and later vehicles have the recommended idle speed printed on a tag somewhere in the engine compartment. For earlier vehicles try about 600 RPM for manual transmissions and 550 RPM in DRIVE for automatics. When setting idle in DRIVE make sure the parking brake is on with the wheels blocked. Do not "flick" the throttle or you could have an out-of-control vehicle.

20. Restricted or dirty air cleaner. Place a 100-watt bulb inside the cleaner element and look at the outside of the element. You can see the dirt pattern. Most times the element can be cleaned with pressurized air. If it looks hopeless, buy a new element.

21. Old and/or cracked spark-plug wires. With the engine idling on a dark night, open the hood and observe the spark-plug wires. If sparks jump around from one wire to another, it is time for a new set. Sometimes the center of carbon-core (TVR) wires can separate. This even happens with new TVR wires. Ignition wires must be removed from the plugs by tugging on the connectors—never by pulling the cables.

22. Leaky intake or exhaust valves. A compression or leak-down test will usually detect this problem; the exception being a hanging or sticking valve that acts up sporadically.

23. Power valve stuck. A piston-type valve can stick open or closed. Diaphragm-type valves seldom stick, but leaking is pos-

sible. A leaky diaphragm causes an abnormally rich condition, evidenced by a black, smoky exhaust and extreme difficulty in restarting.

24. Distributor vacuum-advance line disconnected or leaky. Hoses become dry and crack with age. Connections should be snug because small leaks have great effects.

25. Incorrect basic distributor setting. 1970 and later vehicles have the recommended setting listed on a tag somewhere under the hood. For earlier models, see the service manual.

26. Wrong size main jets. This commonly occurs with racers or enthusiasts and with used carburetors that have passed from hand to hand. Main-jet sizes for most popular Holley carburetors are listed in the Holley Spec Manual.

27. Accelerator-pump failure. Remove the air cleaner and observe the pump nozzles while opening the throttle. A good steady stream of fuel should shoot out of these nozzles as you open the throttle. Make sure you installed the fuel-bowl gasket so the pump passages are not blocked.

28. Clogged idle-transfer slots or holes. This requires a carburetor teardown to find and remedy.

29. Wrong accelerator-pump adjustment. Specific carburetor sections explain the adjustments for your particular carburetor. Repair kit instructions and the Holley Spec Manual give the correct settings.

30. Carburetor too large for the application. This problem occurs usually with the enthusiast. The chapter starting on page 91 deals with this subject.

31. Clogged main jets. This requires internal carburetor inspection. It seldom happens.

32. Hot inlet air system inoperative. This system preheats the inlet air by routing it past the exhaust manifold. It is common on 1969 and later vehicles.

33. Clogged PCV valve. This valve is usually installed in a grommet or cap in the valve cover. Inspection is quick and easy. Most can be cleaned with a little kerosene or solvent. You'll know it's clean when you can blow through it in one direction (toward the carburetor) but not the other. Be sure to install it correctly.

34. Clogged fuel-inlet filter or screen. In some cases this can be inspected without removing the carburetor. Some vehicles

have an in-line filter between the fuel pump and carburetor. Don't hesitate to change it if you are wondering whether you should.

35. Low fuel-pump pressure. A pressure gage can be installed between the fuel pump and carburetor. Typical idle pressures are 3.5 to 6 psi. A pressure below 2.5 psi at top speed is a sure danger signal. Do not plumb fuel lines to a pressure gage inside of your car! Mount a fuel-pressure gage on the cowl outside of the windshield.

36. Throttle plates not reaching wide-open position. All carburetors have a wide-open stop on the throttle lever. Have someone depress the throttle pedal to the floor while you check to see if the throttle lever reaches the wide-open stop. If it does not, a readjustment of the throttle linkage or cable is necessary. Another good idea is to hold the throttle lever against the stop while looking down to see whether the throttle plates are actually in a vertical position.

37. Throttle plates not seated in the bores. This requires removing the throttle body and reseating the plates. This is a rather ticklish operation on most carburetors because throttle-plate screws are staked to prevent loosening. Caution must be taken to prevent twisting off the screw heads. Remove the upset or staked material with a small file and remove the screws with great care. Back off the idle-speed screw, reset the plates in the bores and retighten the screws. Support the opposite side of the shaft and restake with a small prick punch. *Important:* These screws *must* be restaked to avoid screws and/or plates entering the engine to cause serious damage.

38. Dashpot binding or maladjusted. The dashpot is intended to slow the throttle return. If the idle-speed screw never returns to the stop, the dashpot is binding or maladjusted. Normally, the dashpot should have approximately 3/32 additional travel beyond closed throttle. Readjustment can be made by turning the dashpot after the locknut has been loosened. Be sure to tighten the locknut after the readjustment.

39. Vehicle linkage binding. Separate the vehicle throttle linkage from the carburetor as an aid to isolating the problem. Look for sticking or binding along the linkage. A little lubrication may solve the problem. Cable-operated linkages some-times require minor rerouting of the cable to avoid binding.

40. Anti-dieseling solenoid incorrectly adjusted. This solenoid acts as the idle stop while the engine is running and is withdrawn when the ignition is turned off so the plates can close further in the bore to prevent after-run or dieseling. Make sure the solenoid moves back and forth as the ignition is turned on and off. Idle speeds should be set with the solenoid activated (holding throttle slightly away from closed position).

41. Fast-idle cam bound or stuck. Squirting a little solvent on and around the cam and its associated linkage may correct this. A better solution is to remove and clean if time allows.

42. Overstressed throttle-linkage return spring. Check the spring from the carburetor throttle lever to some mounting point on the engine. Separation of the coils usually indicates an overstressed condition and a need to replace the spring.

43. Bound or bent throttle shaft. Separate the throttle-actuating linkage from the carburetor throttle lever and check for binding or sticking. If deposits and dirt are the problem, squirting a little solvent on the shaft may work. If the shaft is bent it will probably have to be replaced. Binding is sometimes caused by over-torquing the carburetor attachment nuts, warping the throttle body. This can prevent diaphragm-operated secondary throttles from ever opening.

44. Worn throttle-shaft bearings. Separate the vehicle throttle linkage from the carburetor and wiggle the throttle shaft up and down. If movement is noticeable, the shaft bearings are worn and the throttle body should be replaced.

45. Failed secondary operating diaphragm. Remove the cover and inspect the diaphragm. Replace the diaphragm if it is torn or broken. Diaphragms are available as a service item and the part number can be obtained from the Holley Carburetor Catalog or the Holley Spec Manual.

46. Plugged secondary vacuum-signal port. This can be repaired by cleaning but requires disassembly of the carburetor.

47. Binding secondary-throttle shaft. This shaft must be extremely free to rotate to function properly. Over-torqued carburetor hold-down nuts or capscrews or the use of a thick soft gasket may result in throttle body warping and hence throttle binding. Another, less-common, cause may be loose or shifted throttle plates.

48. Secondary operating vacuum hose leaky or disconnected. This applies only to 3 x 2 (three-carburetor) applications.

49. Flooding. This is the condition when fuel continues to enter the carburetor even though the float has raised so it should close the inlet valve. Fuel spills out of the carburetor into the intake manifold. The most common causes are leaky or worn fuel-inlet valves or foreign material holding the valve off the seat.

Foreign material is often rust or dirt from the vehicle or gasoline station fuel tank. Flooding can also be caused by a faulty float. Sometimes brass floats develop a leak and lose buoyancy. This hardly ever happens with a closed-cellular float.

50. Broken power valve diaphragm. Remove the power valve and inspect. Excessive backfiring can be the cause.

51. Slow choke come-off. This can be caused by a choke maladjustment, but a more common cause with hot air type chokes is clogging or rupturing of the tube bringing hot air from the stove in the exhaust manifold or crossover to the choke unit. If hot air does not get to the choke bi-metal, the choke will open late or remain in a partially-closed position. When the engine is completely warmed up, the integral-choke unit should be quite hot to the touch.

52. Wrong metering block gasket used with carburetor not having accelerator-pump transfer tube.

53. Model 1920 carburetors. Engines using the crankcase for fuel-vapor storage must be used with a PCV valve in good condition. Because the fuel bowl is vented to the crankcase, high-mileage engines can produce sufficient blow by to pressurize the bowl. This can cause a rich mixture with consequent deterioration of economy and stalling during idle.

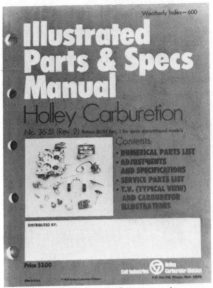

A very valuable book if you work on Holley Carburetors. It contains typical exploded views for most Holley Carburetors. The book is updated yearly and is available from Holley. Your dealer probably has one you can refer to.

You also need a new Holley Performance Parts Catalog every year to stay current with the latest offerings.

Staked choke-plate-retaining screw (arrow). Screws are staked to prevent loosening and falling into the engine. Staking must be removed before taking screws out. Restake with a small punch and hammer on reassembly. Support the shaft while staking to prevent bending the shaft.

Staked throttle-plate screws. Staking prevents accidental loosening of the screws which could cause severe engine damage. If you remove the screws, they must be restaked upon replacement. The shaft must be supported while staking to prevent bending the shaft.

PREPARATION FOR CARBURETOR REPAIR

Once you've analyzed the problem and have decided it lies with the carburetor, you've set yourself up for *another* decision. Here are your options:

1. Take the carburetor apart, make the critical repair and put it back together.
2. Make an economy carburetor repair.
3. Perform a complete carburetor disassembly and repair.

If you've done a little carburetor work and you're fairly sure where the problem lies and time is important, option 1 is probably the best approach. You might be able to perform this operation without removing the carburetor but it isn't really the best idea. Bending over the fender can be difficult and parts are easily lost. If time allows, remove the carburetor. For this option you need a gasket kit at the very minimum. Gasket kit numbers are listed in the Holley Illustrated Parts and Spec Manual. Your dealer probably has one. These manuals contain a wealth of information about specific Holley carburetors and can be obtained by writing:

Aftermarket Sales
Holley Carburetor Division
11955 E. Nine Mile Road
Warren, Michigan 48090

Cost was $3.00 as this book went to press.

It is very seldom necessary to do option 3, disassembling the carburetor and soaking it in a special cleaner as most manuals and printed instructions direct you to do. First of all, it is very time consuming. Most sub-assemblies must be taken apart. Non-metal components cannot be exposed to the cleaner. Commercial carburetor cleaners are expensive, so unless you plan to do several carburetors or are in the business of carburetor rebuilding, this procedure is terribly expensive.

I feel option 2—the economy carburetor-repair method—represents the best compromise of cost, time consumption, and satisfactory results. Cleaners such as kerosene, Stoddard solvent and mild paint thinners allow you to clean up delicate plastic and synthetic parts with metal parts, making complete disassembly unnecessary. A small open can, a brush, a small scraping tool and some elbow grease let you do a fine job of washing. For

dissolving deposits use lacquer thinner, toluol, MEK, Gum-Out, Chem-Tool, etc. These must be used in a well-ventilated area away from fire or pilot lights—such as on a gas range, water heater or furnace. Use with a brush on large surfaces to dissolve deposits. Use a common ear syringe to shoot the solution through passages. Wear safety glasses and keep your hands out of any dissolving-type chemicals. One way or another you can do a good job without exotic equipment.

Passages and orifices should be blown out with compressed air. If compressed air is not available, you can always use a bicycle or tire pump to do the job. Never use a wire or drill to clean orifices and restrictions because the slightest mark can change flow characteristics.

For options 2 and 3, you should obtain the right Pep Kit listed in the Parts and Spec Manual or in the Holley Carburetor Catalog which is also available at the previously-mentioned address for about one dollar. Holley repair kits (Pep Kits) contain the same components as used in the original-equipment carburetor. Not many other repair-kit manufacturers can make this claim.

The Pep Kit contains all gaskets including the one that goes between the carburetor and intake manifold. Also included are fuel-inlet valves, power valves, accelerator-pump diaphragms or cups and a service instruction sheet giving adjustment specifications and procedures.

Parts will probably be left over from your kit when you've finished. Don't panic! The kits were designed to service more than one application in most cases. This "consolidation" reduces the number of kits the dealer has to stock. It is actually more economical to include a few extra parts so each kit will service several different carburetors.

Some special tools are pictured on page 96. These are fine if you plan on working on a lot of carburetors or going into the repair business. Common tools are usually good enough for carburetor disassembly and repair. You will need standard and Phillips screwdrivers, regular and needle-nose pliers and the usual set of open-end wrenches. A sharp scraping tool is a good thing to have around for removing deposits as well as old gaskets from the intake manifold. Be very careful with carburetor sealing surfaces. Zinc and aluminum are easily scratched. A special 1-inch open-end wrench (MAC S-141) is required to remove the fuel-inlet fitting, but you can sometimes make do with a standard open-end wrench. Corbin-clamp pliers and tubing wrenches are good to have for carburetor removal and installation. Tubing wrenches are a *necessity* if you are working with soft fuel-line nuts.

Follow the instructions beginning on page 93 for removing and installing the carburetor. I can't overstress the importance of tagging and identifying each vacuum hose with a piece of tape or shipping tag. Take a few minutes to make a schematic diagram showing all of the hookups. This will save a lot of grief, especially on some later vehicles with complicated emission-control systems.

A holding fixture should be used to prevent damage to the throttle plates while working on the carburetor. Many of these stands or fixtures are available, but 5/16 bolts and nuts work just fine on any carburetor with four mounting holes. A second nut can be used on each bolt to lock the throttle body onto the bolts preventing wobble. Holley's plastic carburetor legs are easiest to use because you just shove them into the holes of the carburetor base.

Now you're ready for the job. Specific carburetor sections include disassembly, repair and assembly procedure for most popular Holley carburetors. In most cases, I have discussed the complete disassembly even though you may be using the economy method. It would be impossible to include every variation of every model carburetor Holley ever made. I picked typical carburetors, but yours may be a little different. I expect you to do a little thinking on your own.

Holley PEP repair kits contain original parts. Many other kits do not and the parts may not perform to the desired specifications.

Two types of bowl vents. One on left used on pre-1968 vehicles vented directly to the atmosphere. Vent on the right used since the advent of evaporative-emission standards connects the bowl to a charcoal canister. It is open during idle and engine off.

Idle-speed screw and anti-dieseling or after-run solenoid. Idle speed should be set to recommended value with solenoid activated.

Contents of a typical 4-barrel PEP kit. Be sure to read the instruction sheet. It contains a lot of tips and most of the adjustment specifications you'll need.

It's a good idea to place the carburetor on a fixture while disassembling to protect the throttle plates. Many good ones are available, but these four 5/16 bolts and eight nuts were purchased for well under a dollar and work just great. We learned this one from Doug Roe.

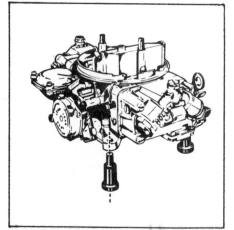

Holley's own plastic carburetor legs snap into throttle-base holes so you can overhaul or adjust a carburetor without danger of throttle-plate damage.

Mass confusion! An excellent example why you should identify all hoses and connections. Make yourself a little sketch for reassembly. Don't rely on memory because it is sure to get you in trouble when you try to reconnect all these hoses.

These are the ordinary kinds of tools you will need for working on carburetors. In addition, tubing wrenches, Corbin-clamp pliers and a special fuel-inlet open-end wrench as shown on page 96 should also be considered as essentials. This same page shows other tools you will find useful as you work on Holley Carburetors.

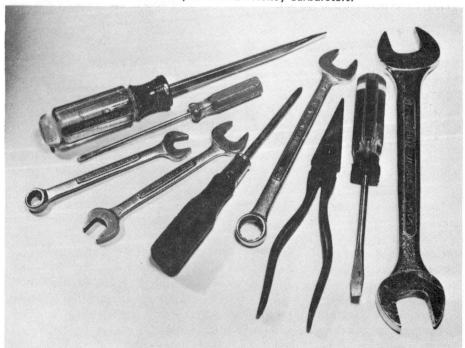

Models 4150/4160, 4165/4175, 2300 Repair & Adjustment

2/Remove the four secondary fuel-bowl screws. Remove fuel bowl and gasket and the metering block and its gasket. Separating these assemblies may require a rap from your screwdriver handle or a rubber or plastic mallet. List 3310-2 carburetors are Model 4160 with a metering plate instead of a metering block. These also tend to stick after the clutch-head screws are removed. Again, a rap should tend to loosen it. The one I used here has dual inlets (Model 4150 style) and no fuel-delivery tube between the bowls.

4/Now repeat the procedure on the primary side. With a Model 2300, two-barrel, this is where you begin. Use plenty of torque when replacing the bowl screws. 50 in. lb. if you're fortunate enough to have a torque wrench with a screwdriver-blade attachment.

The Model 4150 family of carburetors, includes Models 4160, 4165, 4175, and 2300. Because these carburetor models are similar in construction they are discussed in one set of disassembly, repair and adjustment procedures.

For demonstration purposes I used the popular Model 4150 List 3310-1 because it incorporates more sub-assemblies than most carburetors. If you can make it through this one, you won't have any problem with any other carburetor in this family. Exceptions are pointed out.

3/Here is a single-inlet carburetor with the balance tube between the bowls. Note the O-ring (arrow). Don't worry about preserving the O-ring if you are rebuilding—new ones are in the repair kit. Just slide the fuel bowls off of the tube. When reassembling these, use petroleum jelly or oil on the used or new O-rings. Twist tube slightly to ease entry of O-ring into each bowl assembly. Observe carefully to make sure no part of the O-ring gets pinched over the bead of the tube. Fuel under pressure is contained in this tube, so be sure you don't make any mistakes or you could end up with a leak.

5/Next remove the choke unit. With integral chokes, the first thing to do is to remove the little hairpin retainer from the bottom of the rod connecting the choke-control—unit through the main body casting—up to the choke lever. Needle-nose pliers are best for this.

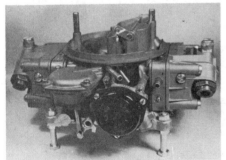

1/Before beginning disassembly, mount the carburetor on a holding fixture or a set of 5/16 bolts, or onto Holley Carburetor Legs. It's a good idea to loosen the fuel-inlet fitting now and also the fuel-bowl sight plugs and needle and seat lock screws because it is easier to do while the carburetor is completely assembled. I prefer to loosen these parts while the carburetor is still securely bolted to the manifold.

6/Before going further, note where the little mark on the black bi-metal housing is relative to the marks on the top of the choke casting. The little strut in the middle of these marks on the casting is called the *index* mark. Choke adjustments are designated relative to this mark. Draw a picture so you can reassemble the choke in the same position. Choke settings are called out in marks rich or lean from the index. Consult the Holley Spec Manual if you've lost or forgotten the setting.

8/The metal choke housing can be removed from the main body by the removal of three attachment screws.

10/Remove clip from shaft (arrow) and take out 3 screws mounting diaphragm housing to main body. Remove diaphragm housing.

7/Remove the three screws holding the bi-metal housing retainer and remove retainer, gasket and bi-metal housing. Do not remove the bi-metal from the housing.

9/For those carburetors with a divorced or remote choke, remove the two screws holding the choke diaphragm, the retainer holding the fast-idle cam and lever and the choke rod retainer if there is one. Disconnect vacuum hose at the throttle body and remove all of the choke parts. This photo shows the choke lever and diaphragm from a typical divorced or remote-choke application.

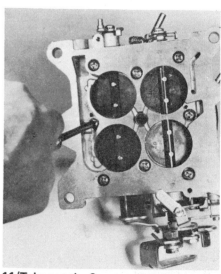

11/Take out the 8 screws attaching throttle body to main body. These are often tough to turn. An impact driver is especially helpful here.

You now have the carburetor broken down to the major sub-assemblies and you can begin their disassembly.

FUEL BOWLS

12/Center-pivot-float fuel bowls hinge the primary floats at the front and secondary floats at the back. This type bowl can have either single or dual inlets. When both sides of the bowl casting are tapped, one side is closed with a plug and the other side has an inlet fitting.

14/Fuel bowl with side-mounted float. Remove baffle (arrow) surrounding inlet valve and then retainer from float hinge pin. Float assembly can then be removed. Non-adjustable inlet valves must be removed from the inside with an open-end wrench. Take out the fuel-inlet fitting you previously loosened. Remove it with the integral filter and spring (see picture below). Remove sight plugs if they are used. You now have the fuel bowl and bowls completely taken apart. This fuel bowl has an externally-adjustable needle/seat.

16/17/Primary bowls (and secondary bowls on double-pumpers) contain accelerator-pump assemblies. Remove four attachment screws and lift pump diaphragm and housing from the bowl. Some have a hanging-ball inlet. Clearance between ball and retainer (arrow) should be 0.011—0.013 inch with bowl inverted. Others have a rubber-umbrella inlet valve (top photo). If you plan to immerse the fuel bowl in a cleaner, the rubber valve must be removed.

13/On bowls with front-mounted floats, first remove the inlet needle-and-seat assembly. Loosen the lock screw and turn the hex nut counterclockwise. The hex nut slips over the needle-and-seat assembly. Loosen the lock screw and turn the hex nut counterclockwise. The hex nut slips over the needle-and-seat assembly. Now remove the two screws holding the float-mounting bracket and remove the bracket and float from the fuel bowl.

15/Disassembled side-mounted-float fuel bowl. This one has a non-adjustable inlet valve.

METERING BLOCKS

18/Remove main metering jets with a wide-blade screwdriver. The screwdriver must cover both sides of the slot or you will damage the jets. Primary and secondary jets are often different sizes. Small jets go in the primary side if the carburetor has smaller primary venturis. On rare occasions jets will differ from side to side in the same metering block. Always check the jet sizes and locations and write this down *before* you take the jets out. Remove bowl-vent splash shield and any vacuum fittings if used. DO NOT REMOVE ANY PRESSED-IN VACUUM TUBES. Take out idle-mixture screws and seals. Remove power valves with a 1-inch wrench. A socket wrench is preferred, but you can do it with an open end if you proceed carefully. The wrench for large fuel-inlet fittings works fine here.

It is not usually necessary to disassemble metering blocks any further. Although there are tubes inside the main and idle wells, these can usually be cleaned with compressed air. The small metering plate from the secondary side of a Model 4160 requires no further disassembly.

Metering blocks with an O-ringed tube connecting them to the main body should have this tube removed before proceeding further.

Use new gaskets on reassembly. If you have to use the existing gaskets, keep them soaking in gasoline or solvent so they will not shrink and/or lose their shape. When you reassemble the carburetor, install the mixture screws 1-1/2 turns off their seats as a starting point.

CHOKE ASSEMBLY

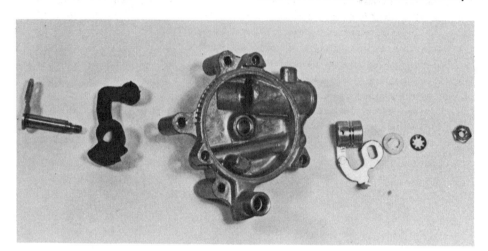

19/20/Take a good look at this assembly before taking it apart. Note the fast-idle cam and choke lever relationship. Remove the choke shaft nut, lock washer and spacer and then slide the shaft and fast idle cam from the housing. Next remove the choke qualifying piston. Make sure it operates freely in its bore. Remove the cork gasket that surrounds the restriction on the back side of the choke housing (arrow). Use a new gasket on reassembly.

SECONDARY DIAPHRAGM ASSEMBLY

21/Remove four attachment screws. A rap will separate the upper and lower housings. The spring and diaphragm can now be removed. My pencil is pointing to a cork gasket that should be removed. Use a new gasket on reassembly.

22/If you plan to immerse the main body in a carburetor cleaner, remove the choke plate and shaft to remove the little plastic guide. If you use the economy repair method, choke-shaft removal is seldom necessary and I don't recommend it. If you plan to remove the choke plate and shaft, first file the stake portion of the choke screws. Then take out the screws and lift the choke plate out of the shaft. The choke shaft can now be removed from the main body allowing the removal of the choke rod and guide. Remember, the choke-plate screws must be restaked upon reassembly. This requires supporting the shaft so it will not be bent. Remove the pump-discharge-nozzle screw, the nozzle and gasket. Turn the body over to remove the pump-discharge valve. Double pumpers have two pump-discharge assemblies. Model 4165 does not have a pump-discharge valve at this location.

CLEANING & ASSEMBLY

Now your carburetor is completely disassembled and ready for cleaning. Remember if you've chosen the complete carburetor repair method, only the metal parts should be immersed in the special cleaner. All non-metal parts including the choke bi-metal and housing, Teflon bushings and the plastic accelerator-pump cams should be cleaned with a milder cleaner such as kerosene. Once all cleaning is done, inspect the parts for undue wear to determine if they should be replaced. If you've purchased the Pep Kit, all gaskets, pump diaphragms, secondary diaphragms (in some cases), fuel-inlet valves and power valves are included.

Blow out all passages and restrictions thoroughly. Make sure all mating surfaces are free of deposits and old gaskets.

24/In reassembling the carburetor, simply follow the disassembly instructions in reverse order. Pictures and exploded views in this book should be very helpful.

THROTTLE BODY

23/It is seldom necessary to disassemble the throttle body and immerse all of the parts. First of all, only a few metering restrictions and small passages are in the throttle body. Secondly, the throttle-plate-attachment screws must be removed and restaked upon assembly. This all adds up to quite a task and if it isn't necessary, why do it? Cleaning with a brush and a milder solvent is usually more than adequate. If you insist on a complete disassembly, your work is going to include: Remove the idle-speed screw and spring, remove the diaphragm-operating lever from the secondary throttle shaft and the fast-idle lever from the primary shaft, remove the cotter key and the connecting link between the primary and secondary throttle levers. File off staked ends of the throttle-plate attachment screws, remove screws and throttle plates. Slide the shafts out of the flange. Take out the Teflon bushings (typical on secondary side of carburetors with diaphragm-operated secondaries).

There are a few things to look for. When assembling the fuel-supply tube, use a little Vaseline on the O-rings so they'll slip in a little easier. When assembled properly, you should be able to rotate the tube with your fingers. Be sure to do a good job of restaking the choke and throttle-attachment screws. If one of these screws comes loose and goes into the engine, it could be very expensive. Other than the very small choke and throttle screws, don't be afraid to torque the assembly screws down, especially the throttle-body and fuel-bowl screws.

Once you've gotten them all in, it's a good idea to go around again and give them a little extra. Be sure that the secondary diaphragm is sealed all the way around when you attach the cover to the lower housing. Unload the diaphragm spring by pushing up on the rod. This will cause the diaphragm to lay flatter. On the Model 4165, the short bowl screws go on top. Make sure the bi-metal loop goes over the choke lever tang when installing the bi-metal housing.

Don't forget to replace all of the gaskets. Some of the little ones are easy to overlook.

25/Some bi-metals have a hooked end while others have a loop as shown. Make sure you capture the tang with the end of the bi-metal. Rotate the bi-metal housing back and forth before tightening the retainer. The choke plate should move as you do so.

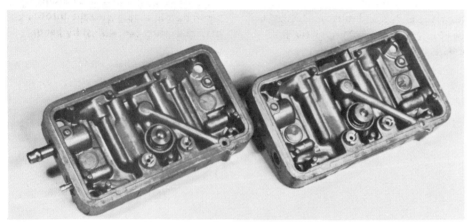

26/The primary metering block can be distinguished from the secondary because it contains the idle-mixture screws and may also contain one or more tubes. Metering-block gaskets are properly installed when they line up with the little dowels on the metering block.

27/Externally-adjustable fuel-bowl float levels can be set on the vehicle. Remove the sight plug. Loosen the lock screw at the top of the assembly and turn the adjusting nut until the fuel level is at the bottom of the sight-plug hole. Flush the bowl a few times by accelerating the engine with the transmission in neutral to confirm your setting. Tighten the lock screw while holding the adjustment nut and replace the sight plug. This operation is difficult to do accurately on a rough-idling vehicle.

28/Dry float setting is usually measured between the bowl casting and the end of the float with the bowl inverted. Holley Pep Kits give the correct spec and gage in some cases. The Holley Spec Manual is another good source. Adjust the float by bending the tab as shown, being careful not to mar the contact surface.

ADJUSTMENTS

A number of adjustments need to be made as you reassemble the carburetor. Adjustment procedures are shown in the accompanying photographs. Specific dimensions are in the instruction sheet supplied with repair kit and in the Holley Spec Manual.

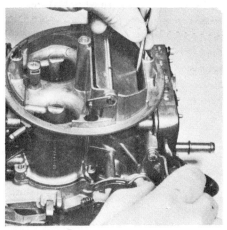

29/Choke-qualify adjustment. Specification is clearance between the choke plate and the casting on either the top or bottom edge as noted. Repair kit or Holley Spec Manual gives this dimension. Hold choke plate closed and measure the clearance with the diaphragm operated by hand or with a vacuum source. A vacuum source is best—your engine is an excellent one. Adjust by bending the diaphragm link on divorced-choke carburetors.

30/Dechoke spec is measured between choke plate and housing with the throttle in the wide-open position. Adjust by bending the end of the choke rod as indicated by screwdriver.

31/Emission-type vent valve connects to charcoal cannister. Clearance should be 0.015 inch as shown with the throttle at normal or curb idle. Adjust by bending lever.

32/Old-style external vent clearance called out in Holley Spec Manual and repair kit is usually around 3/32 inch. Adjust by bending lever.

33/Pump lever should always be capable of at least 0.015 to 0.020-inch additional travel beyond the screw when the throttle is in the wide-open position. Screw and lever should also be in contact at idle. Use screw to accomplish both. This pump cam is in position 2 (arrow). To reduce capacity and change delivery somewhat, remove screw, move cam and insert in hole 1 and screw into alternate hole in cam. With a green cam, added pump-lever travel should be 0.010 inch.

34/Normal or curb idle is set as shown. You'll want to set it to correct engine RPM later, but about 1-1/2 turns from having the plates seated in the bores is a good rough setting.

35/Secondary idle-speed adjustment. About 1/2 to 1 turn away from having the plates seated in the bore is a good approximation.

36/Fast-idle adjustment is made with the fast-idle screw on the highest step of the fast-idle cam. Clearance between the primary throttle plate and the throttle bore should be about 0.025 inch, measured as shown.

37/Dashpot setting refers to the additional travel the dashpot has when the throttle is in the curb-idle position. Consult the Holley Spec Manual for the exact setting. 0.090-0.120 inch is common. Adjust by loosening the locknut and turning the complete dashpot assembly.

WARNING! Model 2300 throttle bores ARE NOT CENTERED. They are offset toward the rear of the carburetor base. If you are careless when installing the gasket, it can prevent correct throttle operation. The engine may start and run, but the throttles may snag in the gasket and not close.

Model 2300 Exploded View
for 3 x 2-barrel setup
Holley Typical View 37–1

**Center carburetor
with accelerator pump
and choke**

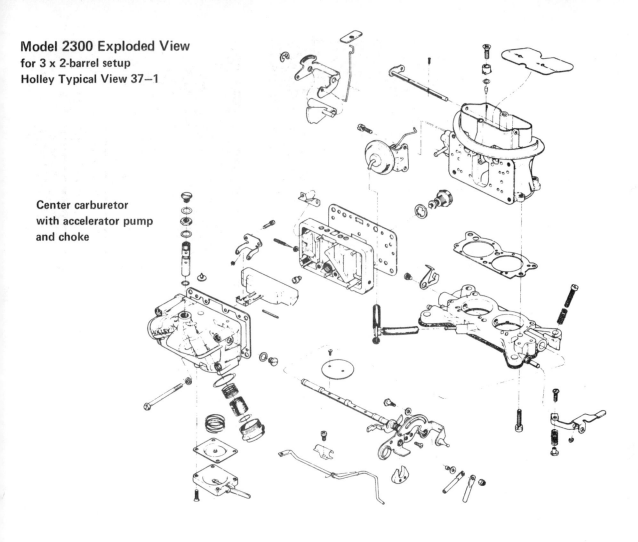

**Front/rear carburetor
with vacuum-actuated
throttles**

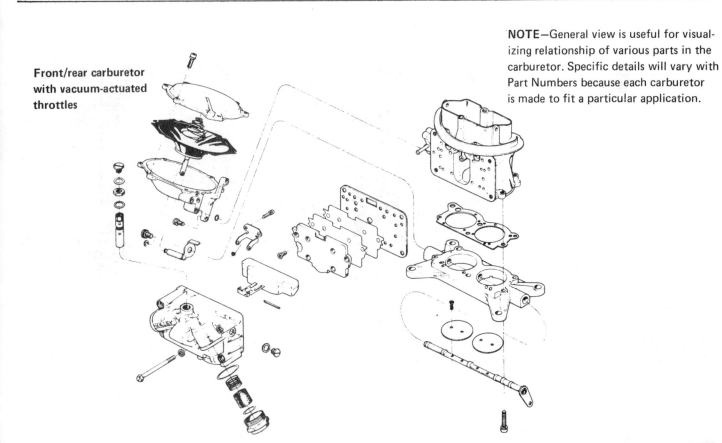

NOTE—General view is useful for visualizing relationship of various parts in the carburetor. Specific details will vary with Part Numbers because each carburetor is made to fit a particular application.

AVOID LEAD POISONING! Tetraethyl lead (Ethyl compound) collects in minute amounts on all used carburetor parts. DO NOT TEST ANY PORTION OF A CARBURETOR BY PLACING YOUR MOUTH AGAINST IT! Continued testing of such components in this manner could result in serious lead poisoning. Carburetor passages can be tested by using air or smoke directed through a piece of tubing (windshield washer or similar).

Some pieces not sold as separate parts — In the exploded views provided throughout this chapter you will see a diamond symbol or the letters "N.S." In either case, this means that the part referenced is not sold as a separate part. If you need one of those pieces — it's new-carburetor time!

1	Choke plate
2	Choke shaft assembly
3	Fast idle pick-up lever
4	Choke hsg. shaft & lev. assy.
5	Fast idle cam lever
6	Choke therm. lev., link & piston
7	Choke plate screw
8	Therm. hsg. clamp screw
9	Throttle stop screw
10	Air vent rod clamp scr. & LW
11	Fuel bowl to main body screw
12	Throt. body scr. & LW
13	Choke hsg. scr. & LW
14	Dashpot brkt. scr. & LW
15	Fast idle cam lever screw
16	Fast idle cam lev. & throt. lev. screw & LW
17	Pump oper. lev. adj. screw
18	Pump discharge nozzle screw
19	Throttle plate screw
20	Fuel pump cov. assy. scr. & LW
21	Pump cam lock scr. & LW
22	Fuel valve seat lock screw
23	Fuel level check plug
24	Fuel level check plug gasket
25	Fuel inlet fitting gasket
26	Fuel bowl screw gasket
27	Choke housing gasket
28	Power valve body gasket
29	Throttle body gasket
30	Choke therm. housing gasket
31	Flange gasket
32	Fuel valve seat adj. nut gskt.
33	Fuel valve seat lock scr. gaskt.
34	Fuel bowl gasket
35	Metering body gasket
36	Pump discharge nozzle gasket
37	Throttle plate
38	Throt. body & shaft assy.
39	Idle adjusting needle
40	Float & hinge assy.
41	Fuel inlet valve & seat assy.
42	Pump oper. lev. adj. scr. fitting
43	Fuel inlet fitting
44	Pump discharge nozzle
45	Main jet
46	Air vent valve
47	Pump discharge needle valve or check ball weight
48	Power valve assembly
49	Fuel valve seat "O" ring seal or gasket
50	Idle needle seal
51	Choke rod seal
52	Choke cold air tube grommet
53	Pump inlet check ball
54	Pump discharge check ball
55	Choke hsg. & plugs assy.
56	Fuel pump cover assy.
57	Fuel bowl & plugs assy.

58	Main metering body & plugs assy.
59	Pump diaphragm assembly
60	Float spring retainer
61	Air vent retainer
62	Fast idle cam lev. scr. spring
63	Throttle stop screw spring
64	Pump diaphragm return spring
65	Fast idle cam lev. spring
66	Pump oper. lev. adj. spring
67	Pump inlet check ball retainer
68	Air vent rod spring
69	Float spring
70	Choke thermostat shaft nut
71	Dashpot screw nut
72	Fuel valve seat adj. nut
73	Choke therm. lever spacer
74	Fast idle cam assembly
75	Pump cam
76	Choke rod
77	Air vent rod
78	Choke therm. shaft nut LW
79	Thermostat housing assembly
80	Choke rod retainer
81	Thermostat housing clamp
82	Dashpot bracket
83	Air vent rod clamp
84	Filter screen assembly
85	Dashpot assembly
86	Baffle plate
87	Pump operating lever
88	Pump operating lev. retainer
89	Adapter mounting & diaphragm cover assy. screw
90	Throt. diaphragm hsg. scr.
91	Adapter passage screw
92	Choke bracket screw
93	Air adapter hole plug
94	Throt. diaphragm hsg. gasket
95	Throttle lever
96	Throttle shaft bearing
97	Throttle shaft brg. (center)
98	Throttle connector pin bushing
99	Diaphragm check ball
100	Throttle connector pin
101	Diaphragm housing cover
102	Air vent cap
103	Diaphragm housing assembly
104	Diaphragm link retainer
105	Air vent rod spg. retainer
106	Diaphragm spring
107	Throttle link connector pin nut
108	Throttle connector bar
109	Choke brkt. scr. lock washer
110	Throt. link connector pin washer
111	Throt. connector pin washer
112	Throttle connector pin spacer
113	Throt. connector pin retainer
114	Choke control lever bracket
115	Metering body vent baffle
116	Throt. diaphragm adapter
117	Diaphragm housing
118	Idle adj. needle spring

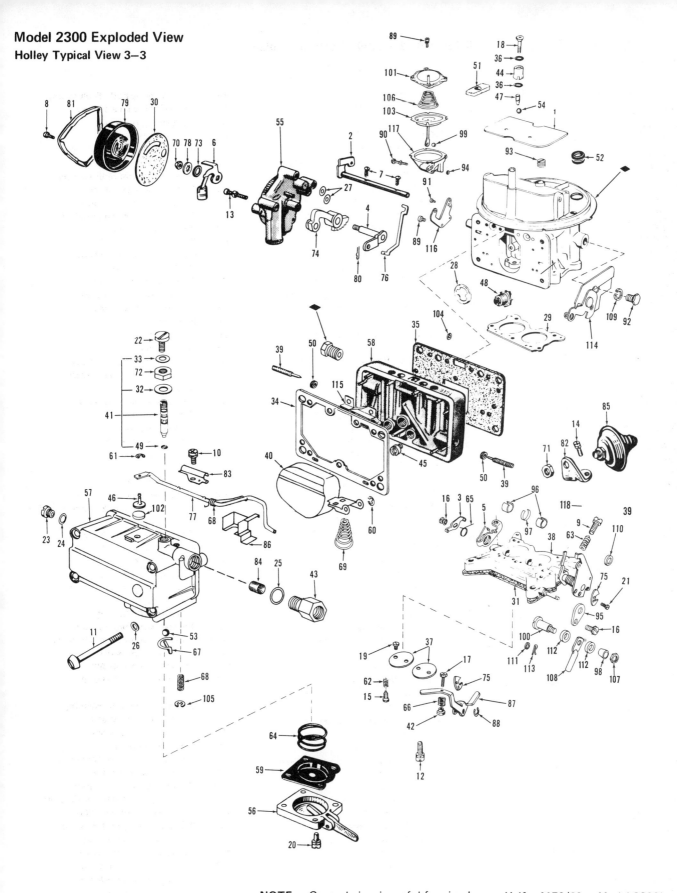

NOTE — General view is useful for visual-izing relationship of various parts in the carburetor. Specific details will vary with Part Numbers because each carburetor is made to fit a particular application.

Half a 4150/60 — Model 2300's are liter-ally the front half of a 4150/60 carbu-retor. Therefore, the systems are as des-cribed for the Model 4150/60 in the following section.

Model 4150/4160 (also 3150/3160) Exploded View Identification

1	Choke plate	61	Flange gasket	128	Choke control lever nut
2	Choke shaft assembly	62	Throttle plate — primary	129	Back-up plate stud nut
3	Fast idle pick-up lever	63	Throttle plate — secondary	130	Throttle lever ball nut
4	Choke housing shaft & lever assy.	64	Throt. body & shaft assembly	131	Dashpot nut
5	Choke control lever	65	Fuel line tube	132	Fuel valve seat adj. nut
6	Fast idle cam lever	66	Balance tube	133	Choke thermostat lever spacer
7	Choke lever & swivel assy.	67	Idle adjusting needle	134	Pump check ball weight
8	Choke therm. lev., link & piston assembly	68	Float & hinge assy. — primary	135	Pump cam
		69	Float & hinge assy. — secondary	136	Fast idle cam assembly
9	Choke rod lev. & bush. assy.	71	Float lever shaft	137	Fast idle cam & shaft assembly
10	Choke plate screw	72	Fuel inlet valve & seat assy.	138	Choke rod
11	Therm. housing clamp screw	73	Pump lever adjusting screw fitting	139	Throttle connecting rod
12	Throttle stop screw	74	Fuel inlet fitting	140	Air vent push rod
13	Choke lever assembly swivel screw	75	Fuel transfer tube fitting assy.	141	Throttle lev. ball nut washer
14	Choke diaph. assy., brkt. scr. & lock washer	76	Fuel inlet tube & fitting assy.	143	Choke shaft nut lock washer
		77	Pump discharge nozzle	144	Choke control lev. nut lock washer
15	Air vent clamp screw & LW	78	Main jet — primary	145	Back-up plate stud nut lock washer
16	Sec. diaph. assy. cov. scr. & LW	81	Air vent valve	146	Throt. connector pin washer
17	Fuel bowl to main body screw — primary	82	Pump discharge needle valve	147	Choke spring washer
		83	Power valve assy. — primary	148	Balance tube washer
18	Fuel bowl to main body screw — secondary	85	Fuel line tube "O" ring seal	149	Therm. hsg. assy. — complete
		86	Balance tube "O" ring seal	150	Throt. connector pin retainer
19	Diaph. lever adjusting screw	87	Fuel valve seat "O" ring seal	151	Choke rod retainer
20	Throt. body screw & lock washer	88	Idle needle seal	152	Throt. connecting rod cotter pin
21	Diaph. hsg. assy. scr. & LW	89	Choke rod seal	153	Choke cont. wire brkt. clamp
22	Choke housing screw & LW	90	Diaphragm housing check ball — sec.	154	Thermostat housing clamp
23	Dashpot brkt. screw & LW	91	Pump inlet check ball	155	Choke control wire bracket
24	Fast idle cam lever adj. screw	92	Throttle lever ball	156	Dashpot bracket
25	Fast idle cam lev. scr. & LW	93	Pump discharge check ball	157	Air vent rod clamp
26	Diaph. lev. assy. scr. & LW	94	Choke diaphragm assembly link	158	Fast idle cam plunger
27	Throt. plate screw — primary	95	Sec. diaphragm housing cover	159	Choke vacuum tube
28	Throt. plate screw — secondary	96	Back-up plate & stud assembly	160	Fuel transfer tube
29	Pump lever adjusting screw	97	Fast idle cam plate	161	Filter screen
30	Pump discharge nozzle screw	98	Secondary metering body plate	162	Dashpot assembly
31	Fast idle cam plate scr. & LW	99	Air vent cap	163	Baffle plate — primary (brass)
32	Choke cont. wire brkt. clamp scr.	100	Choke hsg. & plugs assembly	164	Baffle plate — secondary
33	Pump cam lock screw	101	Main metering body & plugs assy. — primary	165	Metering body vent baffle
34	Fuel pump cov. assy. scr. & LW			166	Float shaft retainer bracket
35	Secondary metering body screw	102	Main metering body & plugs assy — secondary	167	Fuel inlet filter
36	Throt. body screw — special			168	Diaphragm lever assembly
37	Fuel valve seat lock screw	103	Fuel pump cover assembly	169	Pump operating lever
38	Float shaft brkt. scr. & LW	104	Fuel bowl & plugs assy. — primary	170	Pump operating lever retainer
39	Spark hole plug	105	Fuel bowl & plugs assy. — secondary	171	Secondary diaphragm housing
40	Fuel bowl plug	106	Secondary diaph. & rod assy.	172	Throt. shaft bearing — sec (ribbon)
41	Fuel level check plug	107	Pump diaphragm assembly	173	Throt. shaft bearing — sec. (ribbon)
42	Fuel level check plug gasket	108	Choke diaphragm assembly — complete	174	Throt. shaft bearing — pri. (solid)
43	Fuel inlet fitting gasket	109	Secondary diaph. link retainer	175	Fuel bowl drain plug
44	Fuel valve seat gasket	110	Air vent rod spring retainer	176	Choke assy. — complete (divorced)
45	Fuel bowl screw gasket	111	Float retainer	177	Choke shaft lever
46	Sec. diaphragm housing gasket	112	Air vent valve retainer	178	Idle by-pass adj. screw
47	Choke housing gasket	113	Choke control lever retainer	179	Choke lever screw & LW
48	Power valve body gasket	114	Fast idle cam plunger spring	180	Choke piston link retainer
49	Choke thermostat housing gasket	115	Fast idle cam lever screw spring	181	Fast idle cam retainer
50	Sec. metering body plate gasket	116	Throttle stop screw spring	182	Choke rod clevis clip
51	Fuel valve seat adj. nut gasket	117	Secondary diaphragm spring	183	Idle by-pass adj. screw spring
52	Fuel valve seat lock screw gasket	118	Diaphragm return spring	184	Choke piston lever spacer
53	Throt. body screw gasket	119	Fast idle cam lever spring	185	Choke piston & link assembly
54	Pump discharge nozzle gasket	120	Pump lev. adj. screw spring	186	Choke oper. lev. spring washer
55	Metering body gasket — primary	121	Air vent rod spring	187	Choke oper. lever washer
56	Metering body gasket — secondary	122	Pump inlet check ball ret. spring	188	Choke clevis pin
57	Fuel bowl gasket	123	Choke spring	189	Baffle plate — primary (plastic)
58	Throttle body gasket	124	Float spring — pri. & sec.		
59	Fuel bowl plug gasket	125	Fuel inlet filter spring		
60	Fuel inlet filter gasket	126	Choke cont. wire brkt. clamp scr. nut		
		127	Choke thermostat shaft nut		

Model 4150/4160 Exploded View
(also 3150/3160)
Holley Typical View 24—1

CARBURETOR PART NUMBER
IS ON AIR HORN FLANGE

LIST 1971

END INLET MORAINE FILTER BOWL

VARIABLE PARTS FOR LIST
NO. R—1971A & R—2052A

CENTER INLET FUEL BOWL

NOTE — General view is useful for visualizing relationship of various parts in the carburetor. Specific details will vary with Part Numbers because each carburetor is made to fit a particular application.

PLASTIC

BRASS

N.S.

177

Model 4165 Exploded View Identification

1	Choke plate	57	Pump lever adj. screw fitting	114	Secondary connecting rod washer
2	Choke shaft assembly	58	Fuel inlet fitting	115	Throttle seal washer
3	Fast idle pick-up lever	59	Pump discharge nozzle - primary	116	Choke spring washer
4	Choke control lever	60	Pump discharge nozzle - secondary	117	Secondary connecting rod cotter pin
5	Fast idle cam lever	61	Main jet — primary	118	Choke wire bracket clamp
6	Choke lever & swivel assembly	62	Main jet — secondary	119	Choke wire bracket
7	Choke rod lever & bushing assembly	63	Pump check valve	120	Fast idle cam plunger
8	Choke plate screw	64	Power valve assy - primary	121	Choke vacuum hose
9	Choke lever swivel screw	65	Power valve assy — secondary	122	Metering body vent baffle — pri. & sec.
10	Choke diaphragm bracket screw & LW	66	Fuel line tube "O" ring seal	123	Float shaft retaining bracket
11	Fuel pump cover screw & LW — primary	67	Fuel valve seat "O" ring seal	124	Baffle plate — primary
12	Fuel pump cover screw & LW — secondary	68	Idle needle seal	125	Baffle plate — secondary
13	Fuel bowl screw (long) pri. & sec.	69	Choke rod seal	126	Fuel inlet filter
14	Pump lever adjusting screw — secondary	70	Choke diaphragm link	127	Pump operating lever — primary
15	Throttle body screw & LW	71	Back-up plate & stud assy.	128	Pump operating lever & guide assy. — sec.
16	Fast idle cam lever adj. screw	72	Fast idle cam plate	129	Pump cam lever — secondary
17	Fast idle cam lever screw & LW	73	Fuel pump cover assy. - primary	130	Pump operating lever retainer — pri. & sec.
18	Pump cam lever screw & LW	74	Fuel pump cover assy — secondary	131	Choke therm. lever
19	Pump lever adj. screw — primary	75	Fuel bowl & plugs assy — primary	132	Choke therm. cover screw
20	Fast idle cam plate screw & LW	76	Fuel bowl & plugs assy — secondary	133	Choke housing screw
21	Throttle plate screw — pri. & sec.	77	Metering body & plugs assy — primary	134	Choke housing gasket
22	Choke wire bracket clamp screw	78	Metering body & plugs assy — secondary	135	Choke therm. cover gasket
23	Pump cam screw	79	Pump diaphragm assy. — primary	136	Tube & "O" ring assy.
24	Pump discharge nozzle screw	80	Pump diaphragm assy. — secondary	137	Idle adj. needle limiter cap
25	Fuel valve seat lock screw	81	Choke diaphragm assy.	138	Choke housing & plugs assy.
26	Pump operating lever adj. screw	82	Float hinge retainer	139	Choke therm. cover retainer
27	Float shaft bracket screw & LW	83	Cam follower lever assy. retainer	140	Choke therm. shaft nut
28	Throttle stop screw	84	Choke control lever retainer	141	Choke shaft nut lock washer
29	Fuel bowl screw (short) pri. & sec.	85	Pump lever stud	142	Choke housing screw & L.W.
30	Throttle lever extension screw	86	Fast idle cam plunger spring	143	Choke therm. cover assy.
30A	Throttle body channel plug	87	Fast idle cam lever screw spring	144	Choke shaft spacer
31	Fuel level check plug	88	Throttle stop screw spring	145	Cam follower stud
32	Vacuum tube plug	89	Diaphragm return spring — primary	146	Secondary diaphragm cover screws
33	Fuel level check plug gasket	90	Choke spring	147	Secondary diaphragm housing screws
34	Fuel bowl screw gasket	91	Float spring — secondary	148	Secondary metering body screws
35	Power valve gasket	92	Float spring — primary	149	Secondary diaphragm housing gasket
36	Fuel valve seat adj. nut gasket	93	Pump lever adj. screw spring — primary	150	Secondary metering body plate gasket
37	Fuel valve seat lock screw gasket	94	Pump lever adj. screw spring — secondary	151	Tube & "O" ring assy.
38	Pump discharge nozzle gasket	95	Fuel inlet filter spring	152	Four-way connector
39	Metering body gasket — pri. & sec.	96	Diaphragm return spring — secondary	153	Diaphragm cover machine
40	Fuel inlet filter gasket	97	Throttle return spring — secondary	154	Secondary metering body plate
41	Fuel inlet fitting gasket	98	Fast idle cam lever spring	155	Secondary check valve
42	Fuel valve gasket — pri. & sec.	99	Choke wire bracket clamp screw nut	156	Secondary diaphragm
43	Throttle body gasket	100	Back-up plate stud nut	157	Secondary diaphragm link retainer
44	Flange gasket	101	Choke lever nut	158	Secondary diaphragm spring
45	Fuel bowl gasket - pri & sec.	102	Fuel valve seat adj. nut	159	Choke vacuum hose
46	Throttle plate — secondary	103	Pump operating lever adj. nut	160	Choke vacuum hose
47	Throttle plate — primary	104	Throttle lever ext. screw nut	161	Secondary housing & seat assy.
48	Throttle lever extension	105	Pump cam — primary	162	Vent valve screws
49	Cam follower lever assy.	106	Fast idle cam & shaft assy.	163	Air vent rod clamp screw & L.W.
50	Throttle body & shaft assy.	107	Pump cam — secondary	164	Vent valve body assy.
51	Fuel line tubing	108	Fast idle cam assy.	165	Vent valve spring
52	Idle adjusting needle	109	Pump operating lever screw sleeve	166	Vent rod spring
53	Float & hinge assy. — primary	110	Choke rod	167	Vent valve rod
54	Float & hinge assy. — secondary	111	Secondary connecting rod	168	Vent valve clamp assy.
55	Float shaft	112	Back-up plate stud nut LW	169	Air vent rod clamp
56	Fuel inlet valve & seat assy.	113	Choke control lever nut LW	170	Throttle body spacer

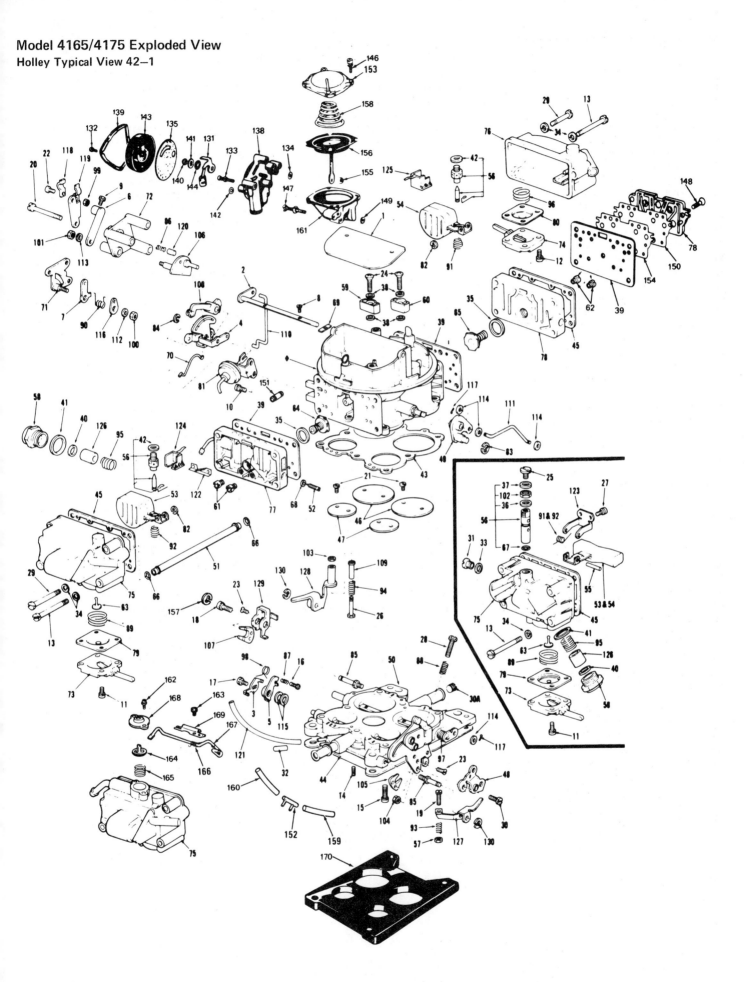

FIRE–A Hazard You Can Minimize

By Howard Fisher

Carburetors and fuel-supply system components carry or contain gasoline—a very flammable liquid. Care in installation and observation of what is happening when fuel pressure is applied to the carburetor (watching for flooding, etc.) can go a long way towards eliminating fire. Where a car or a carburetor is being so worked on in a closed area containing a flame—such as the pilot light or burner of a water heater or furnace—fire danger is extreme. It just takes one spill to generate sufficient vapor to be carried across the floor to the flame. Then the trouble begins in the form of an explosion or a fire or both. Remember that a carburetor removed from a car contains gasoline and this should be drained before carrying the carburetor inside to work on it. This is especially true when you work in an area which has any kind of open flame.

"75% of all auto fires directly result from leaving off the air cleaner," say M. L. Wikre and D. C. Whiting of the Los Altos, California, Fire Department. "Fuel spews out of the carburetor onto the manifold—or standoff collects on the underside of the hood—then a backfire ignites the fuel. The owner usually tells us that he left the air cleaner off, because he had been tuning the car." These fire experts point out that the air cleaner prevents a backfire from igniting any stray fuel. The air cleaner itself does not support combustion very well. It's interesting to note that flame arresters are required on the carburetors of all marine inboard gasoline engines. If a fire starts in the air cleaner, a lot of smoke can be expected, but not much burns if the element is the usual paper type. Nearly all of the auto fires not caused by leaving the

air cleaner off seem to be caused by a bad fuel line or connection between the fuel pump and carburetor.

If a fire extinguisher is carried in the car, the fire can usually be put out quickly and with little damage. A 2-1/2-pound **ABC** dry-chemical extinguisher covers all three classes of fire found in cars. **A**—upholstery and the interior are common burnables, **B** refers to gasoline/oil fires, **C** is an electrical fire.

These extinguishers are small, lightweight and virtually fool-proof. Volume-for-volume, they have much better extinguishing capabilities than carbon dioxide (CO_2). Also, powder retards the fire from flashing up again at hot spots and wiring.

Drawbacks of the dry-chemical extinguisher are that the powder goes *everywhere,* making a mess which must be cleaned up afterward. And, it can go into the engine, especially if there is no air cleaner. If a lot of powder has to be directed into the carburetor air inlet, some will actually get into one or more cylinders through an open intake valve. The piston will compress this into a cake which may prevent the engine from turning when you attempt to restart. Even if you can crank the engine over, the powder is abrasive. So, if very much gets in the engine, pull the cylinder heads and clean out the powder before running the engine again.

CO_2 extinguishers are preferred by many because they cannot damage the engine and there is no after-mess to clean up. This is their greatest *plus*. But, because CO_2 fights fire by replacing the oxygen, a much larger extinguisher (than dry chemical) is required to match the power of a dry-chemical unit. The

fire experts we talked with recommended a 50-pound CO_2 unit! In open areas, especially when the wind is blowing, CO_2 dissipates very quickly and sometimes will not put out a burning fuel line or wire.

If you have a CO_2 and a dry-chemical extinguisher on hand always use the CO_2 first. If that gets the fire out, there's no mess to clean up. For that reason, a large CO_2 extinguisher is the thing to have in your garage—*if you can afford it.* CO_2 extinguishers require more upkeep than a dry-chemical type and they should be professionally checked and recharged every year.

If you have a fire under the hood, don't throw the hood open because hot gases and flames will rush out. Open the hood just a crack and shoot the extinguishing media in. Better yet, shoot it in from under the engine.

Check everything before restarting the engine or you could start a worse fire—just when your extinguisher is all used up. Check all of the ignition wires and the fuel lines to see whether they have burned or melted. Check the fuel line connections to make sure that they have not loosened.

Professional racers are turning to on-board extinguisher systems which pipe pressurized FE1301 Freon to all critical areas. While the first cost of these may seem expensive, the added margin of safety, which these give the driver who may have to exit from a flaming vehicle, certainly offsets their seemingly high first cost. FE1301 extinguishes about three times better than CO_2. The Freon dissipates afterward with no mess and no harm to the engine or other components.

Model 1920
Repair & Adjustment

This is the simplest of the Holley carburetors and it is the easiest to repair. First, set the carburetor on a repair stand. You can use the 5/16 bolts and nuts like we used on the four-bolt-flange carburetors if you hold two bolts in a vise and use two nuts on each bolt so they "capture" the carburetor base. You can get by without a stand if you are very careful. Hold the carburetor in one hand as you disassemble it with the other, being careful not to bump the throttle plate which is wide open on most of these carburetors.

1/Remove vacuum hose V from the choke diaphragm to the throttle portion of the main body. Now remove two screws holding the choke diaphragm bracket to main body. Remove diaphragm and link connecting it to the choke lever. The vent-valve lever and spring are also held by the diaphragm bracket. Remove fuel-inlet fitting from main body with a 5/8-inch wrench. Remove and discard gasket between fitting and main body. Inlet needle and seat F are an integral part of this fuel-inlet fitting.

2/Remove four screws holding fuel bowl to main body and remove bowl. Remove bumper spring 1 from beneath the float and slide baffle from fuel bowl. Discard the gasket. Carefully remove the E-ring 2 retaining the float and slide the float from the pivot or fulcrum pin. On reassembly, tighten screws snugly or to 30 in. lb. if you have a torque wrench. Remove five screws holding metering block to main body. Two shorter screws go into the upper-right and lower-left holes as you face the fuel-bowl side. Now remove the metering block from the main body with a little coaxing or rapping with your screwdriver handle.

3/Power valve or economizer (arrow) is removed by taking out three screws retaining its cover. Lift the cover and the complete power-valve assembly. The one gasket which goes between the diaphragm guide stamping and the carburetor main body is included in the repair kit with a new power-valve-diaphragm assembly. Accelerator-pump diaphragm is visible inside fuel bowl. Metering block is in foreground of this photo.

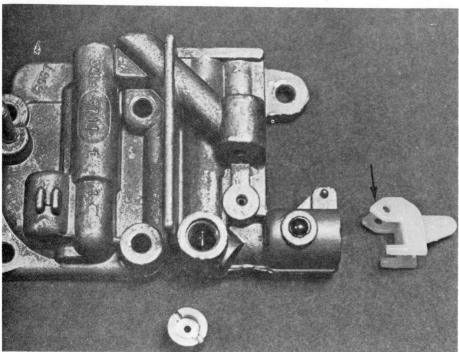

4/The main-metering jet can be removed from the metering body with a wide-blade screwdriver. The blade must be wide enough to cover both slots. If you plan to immerse the metering body into a commercial cleaner, remove the plastic power-valve lever by spreading the ears carefully and sliding them over the pivot pin. Do not remove any other pressed-in restrictions or plugs from the metering body.

5/Remove the retainer and the pump-operating lever from the fulcrum pin. This plastic lever must not be immersed in a harsh cleaner. Next, remove the fast-idle-cam screw, cam, and link connecting it to the choke lever. Remove retainer from link connecting pump-actuating lever to throttle and remove link. Rotate pump lever clockwise as far as it will go. This allows sliding the pump diaphragm and stem assembly and spring from the main body. A new diaphragm is included in the kit.

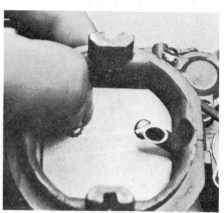

6/Turn idle-mixture-screw limiting cap clockwise until it reaches its stop. Remove cap with a pair of pliers without rotating or bending the screw. Turn the screw in by hand until it reaches the seat, counting the turns as you go. Now remove the screw and spring. When you reassemble the mixture screw, turn it to the same setting and slide a new limiting cap onto the screw with its tang against the lower side of the stop on the main body. If your carburetor does not have a limiting cap, count the turns off the seat so you can start off with a nearly correct idle mixture.

If the throttle bore area is very dirty and the shaft is binding, you may want to remove the throttle shaft and plate. This requires filing off the staked material from the back of the retaining screws. Restaking upon reassembly must be done, even though these screws have lock washers. This is time consuming and touchy, so don't do it if you don't have to. When reinstalling the plate, back off the idle-speed screw and start the screws into the shaft. Hold the carburetor up to the light and seat the plate so it fits well all around the bore.

7/The choke plate and shaft can be removed and replaced without worrying about restaking because star washers are used to lock the attachment screws on most of these carburetors. Remove the choke plate and shaft if you think it warrants it. When reinstalling the choke plate, make sure it fits inside the air horn before tightening the attachment screws. Then check it for freedom of movement.

8/Remove two screws holding bowl-vent cover and seat to the air horn. Earlier versions do not have this vent. Remove valve retainer, valve and spring from main body. Photo shows vent valve disassembled. This vent connects to the charcoal canister by a tube (arrow). Earlier carburetors had mechanical vents to the atmosphere.

9/Check choke diaphragm for leaks by depressing stem and holding your finger over the end of the vacuum tube. If stem moves more than 1/16 inch in ten seconds, replace the diaphragm unit. Now the carburetor is completely disassembled and ready for cleaning and inspection. Remember: Only metal parts should be immersed into commercial carburetor cleaners. Use milder solvents such as kerosene for the remaining pieces.

After cleaning, inspect all parts for wear. If the throttle-shaft bearings are badly worn the carburetor will have to be replaced. This is the exception rather than the rule. Check the idle-mixture needle or screw for a groove on the taper. Replace it if it is badly grooved. Blow out all passages and restrictions with compressed air. A bicycle pump will do nicely. Never stick wires or drills into metering restrictions.

Put the carburetor back together following the disassembly procedures in reverse order. Apply about 30 in. lb. of torque to the fuel bowl screws for good sealing. If you don't have a torque wrench, make them snug.

Several adjustments should be made while the carburetor is off the car. The float must be set before replacing the fuel bowl. Other adjustments are the choke-kick or qualify, rough fast and curb idle, fast-idle-cam position, vent setting and the dashpot adjustment if there is one. The accompanying photographs will help you make these adjustments.

10/Check dry float level. Invert carburetor and measure clearance between the float and casting at location shown. Setting is in the repair kit and the Holley Spec Manual. Adjust by bending the tab on the float lever. Make sure you have bumper spring (arrow) in place when you reassemble the carburetor.

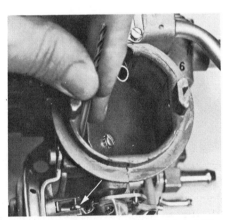

11/Choke qualify is checked with a drill between the choke plate and casting as shown. Activate the choke diaphragm with a vacuum source, put enough pressure on the plate with your finger to move the modulator (arrow) to its stop and take the measurement. Specification is in the repair kit or the Holley Spec Manual. Adjust by bending link.

12/With the fast-idle screw against the highest step of the fast-idle cam, 0.028-inch clearance between the throttle plate and bore yields a rough fast-idle speed setting. You'll reset fast-idle speed on the vehicle later. See your manufacturer's service manual or use about 1700 RPM in neutral with a warm engine with the fast-idle screw on the top step of the cam.

183

13/Normal or curb-idle setting. Two turns from where the throttle plates are seated in the bore is a rough setting. Reset idle to specification when you have the carburetor on the car.

14/With the fast-idle screw and fast-idle cam set as shown, clearance between casting and plate should be set to correct spec, to assure proper fast-idle-cam phasing. Adjust by bending the link (arrow). If you don't have a service manual, 0.100 inch is a good approximation.

15/Correct installation of bowl-vent lever and spring.

NOTE: Crankcase blowby is caused by normal engine wear and/or poor maintenance. The PCV valve and breather cap must be clean and operating at all times. Excessive blowby can pressurize the carburetor bowl via the vent hose, causing rough idle, stalling, poor drivability and poor fuel economy. This can be diagnosed by removing the 3/4-inch vent hose from the top of the carburetor fuel bowl. If the idle smooths out and fuel economy improves, engine condition and/or emission controls—**NOT THE CARBURETOR**—is the problem.

16/Clearance between lever and vent valve should be 0.015 inch with throttle at curb idle. Old-style, pre-emission, atmospheric vent (not shown) is set at 0.090 inch at curb idle. Adjust by bending the lever.

Model 1920 Exploded View Identification

1	Automatic choke assembly
2	Choke plate assembly
3	Choke shaft & lever assy.
4	Choke lever
5	Choke control lever
6	Choke control lev. swivel scr.
7	Main well & economizer body scr. & LW (short)
8	Main well & economizer body scr. & LW (long)
9	Fuel bowl scr. & LW
10	Throt. plate scr. & LW
11	Choke plate scr. & LW
12	Econ. body scr. & LW
13	Wire clamp scr. & LW
14	Therm. cover clamp scr. & LW
15	Fast idle cam screw
16	Choke lever screw
17	Mechanical vent scr. & LW
18	Throt. adapter scr. & LW
19	Dashpot bracket screw
20	Choke brkt. retainer screw
21	Wire bracket screw
22	Choke piston plug
23	Fuel inlet seat gasket
24	Economizer body gasket
25	Float bowl gasket
26	Flange gasket
27	Therm. housing gasket
28	Throttle plate
29	Throt. lever & shaft assy.
30	Throt. lever drive assy.
31	Idle adjusting needle
32	Float & lever assy.
33	Fuel inlet needle & seat assy.
34	Main jet
35	Mechanical vent valve
36	Economizer stem assembly
37	Economizer body cover assy.
38	Fuel inlet seat "O" ring seal
39	Diaphragm push rod sleeve ball
40	Pump operating link
41	Choke piston link
42	Choke diaphragm link
43	Float bowl
44	Main well & econ. body & plugs assy.
45	Pump diaphragm & rod assy.
46	Choke diaphragm assy. — complete
47	Float spring retainer
48	Mechanical vent spring retainer
49	Choke diaphragm link retainer
50	Idle adjusting needle spring
51	Pump return spring
52	Pump operating spring
53	Mechanical vent spring
54	Float spring
55	Choke lever spring
56	Wire clamp screw nut

Model 1920 Exploded View
Holley Typical View 29-2

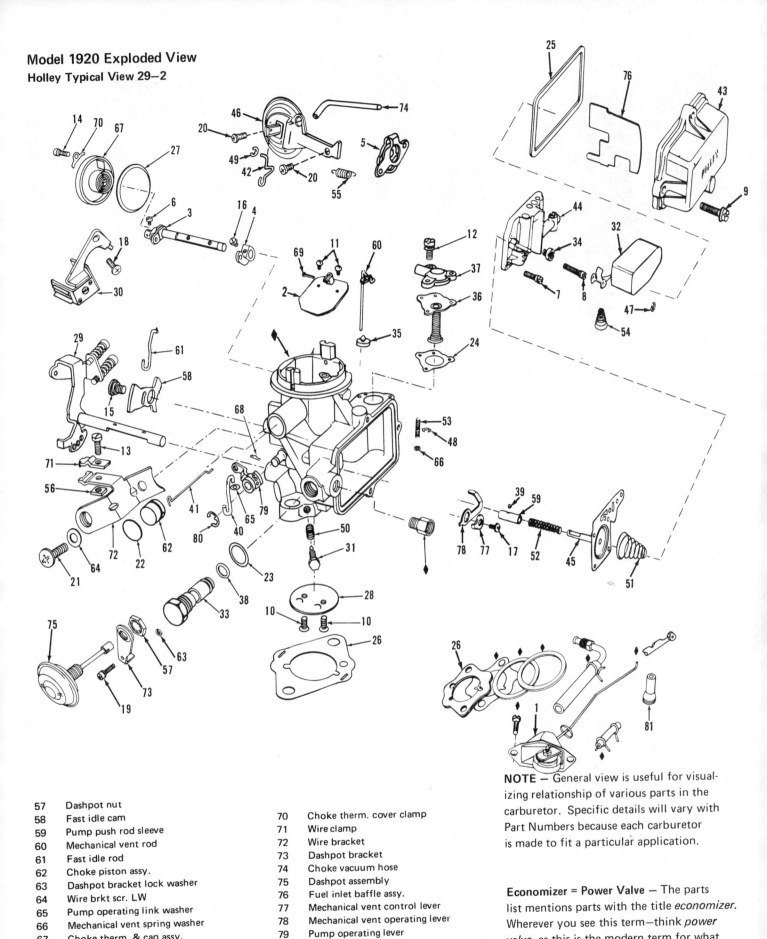

57	Dashpot nut	70	Choke therm. cover clamp
58	Fast idle cam	71	Wire clamp
59	Pump push rod sleeve	72	Wire bracket
60	Mechanical vent rod	73	Dashpot bracket
61	Fast idle rod	74	Choke vacuum hose
62	Choke piston assy.	75	Dashpot assembly
63	Dashpot bracket lock washer	76	Fuel inlet baffle assy.
64	Wire brkt scr. LW	77	Mechanical vent control lever
65	Pump operating link washer	78	Mechanical vent operating lever
66	Mechanical vent spring washer	79	Pump operating lever
67	Choke therm. & cap assy.	80	Pump operating lever retainer
68	Pump operating link retainer	81	Throttle lever bushing
69	Choke piston link retainer pin		

NOTE — General view is useful for visualizing relationship of various parts in the carburetor. Specific details will vary with Part Numbers because each carburetor is made to fit a particular application.

Economizer = Power Valve — The parts list mentions parts with the title *economizer.* Wherever you see this term—think *power valve,* as this is the modern term for what was once called the *economizer valve.*

185

1/These two carburetors are very similar. Model 1940 at left with an integral choke is used for Ford engines. On the right is a Model 1945 used on Chrysler original equipment. It has a remote or divorced choke.

Models 1940 & 1945 Repair & Adjustment

Set the carburetor on a repair stand. You can use the 5/16 bolts if you hold the bolts in a vise and use two nuts on each bolt so as to support the carburetor base. You can get by without a stand if you are especially careful not to damage the throttle plates.

2/Disconnect the vacuum hose at the throttle body and the choke diaphragm then remove two screws holding the diaphragm bracket so you can remove diaphragm and bracket assembly and link connecting it to the choke lever.

3/Now remove fast-idle-cam-retaining clip (arrow) or screw, along with cam and link. Remove dashpot if there is one.

4/Remove seven screws holding air horn or fuel-bowl cover to main body and lift air horn straight up until accelerator pump, main well tube and power valve stem clear the main body. A rap with a plastic hammer or screwdriver handle may be necessary to separate these two castings. Be careful—do not bend the main-well tube. Twist air horn to free pump lever from pump rod. Get screws tight upon reassembly. Use 30 in. lb. if you have a torque wrench.

5/Remove gasket and discard. Do not use a scraper to remove gasket material from air-horn sealing surface because you could destroy sealing bead (arrow). This air horn has a mechanical over-ride on the vacuum power valve (arrow).

6/Remove the pump-operating-rod retainer and screw from the air horn. Disconnect pump stem from the operating rod and remove the pump stem, cup and spring. Most repair kits include a new pump cup. Rotate pump-operating rod and remove it from air-horn casting. Remove pump-rod grommet from air horn. A little Vaseline or silicone grease will aid in inserting the new grommet upon reassembly. This photo shows disassembled air horn. The main-well tube should never be removed unless absolutely necessary. A spring that goes under the power-valve piston is not shown here. If the choke shaft is binding or the plate is sticky or bent, you'll have to remove them. If they appear in good condition don't bother. When removing the choke plate, first remove the staking from the screw threads before taking out the screws. These screws must be restaked on reassembly to prevent their coming loose and falling into the engine. Support the choke shaft while performing this operation to prevent bending the shaft. Also, seat the choke plate in the air horn before torquing the attachment screws.

7/Remove the fuel-inlet fitting (arrow) and gasket from the main body with a 5/8 wrench. This fuel fitting also contains the inlet valve and baffle. A new one is included in the repair kit. Letters identify jets: Main jet (M), power valve (P).

5A/This is a modulated power valve. "Modulated" means the piston moves gradually as manifold vacuum changes. For instance, the power valve might open between 10 and 4 inches of mercury manifold vacuum instead of between 6 and 4 inches with a more conventional valve. Arrow indicates staking around piston retainer. If the power-valve vacuum piston is working freely, leave it alone. If it is binding you'll have to remove it. First remove the staked material holding the piston retainer using the tip of a small file. Remove the piston, retainer and spring. The retainer must be restaked upon reassembly with a small screwdriver, punch or chisel.

8/Remove the spring retainer, hinge pin and float from fuel bowl. Remove main-metering jet (M), using a screwdriver with a blade wide enough to cover the slot in the jet. Carefully depress the power-valve stem and remove power valve (P), seat and spring. Jets shown in photo at top of page.

9/Turn the assembly up-side-down and take out the pump discharge ball and weight. There are two pieces and you need both when you reassemble the carburetor. Remove three screws holding throttle body to main body and separate the two castings. Discard the gasket. The main body is now completely disassembled.

10/Turn idle limiter (arrow) clockwise as far as it will go. This is the leannest position. Remove the limiter cap with a pair of pliers, trying not to rotate or bend the mixture screw. Turn the screw with your fingers until it seats, counting the turns as you go. Record this setting for future reference. Now remove screw and spring. If your repair kit contains a limiter cap, set mixture screw to the reference setting and install cap in the leannest position when assembling. Remove main-body to throttle-body gasket. This is usually as far as you'll have to go with the throttle body.

11/If the throttle bore is very dirty or the throttle plate is damaged or binding, completely disassemble the throttle body. Remove nut and lock washer from end of throttle shaft. Then release tension from positive-return spring by unhooking inside tang. Slide complete positive-return-spring assembly from the shaft. Note how the individual parts fit together. Refer to the photograph.

12/Next, remove staking from the throttle-attachment screws and withdraw the screws carefully. The throttle lever and shaft assembly can now be removed. The carburetor is completely disassembled. The throttle-attachment screws must be restaked and the throttle plate reseated in the bore when you reassemble the unit. Photo shows positive-return components in their relative positions.

The carburetor is now ready for cleaning. Remember: Only metal parts can be immersed in the commercial carburetor cleaner. Non-metal parts must be cleaned with one of the milder solvents mentioned in the repair-preparation section. Make sure all old gasket material is removed from sealing surfaces. While cleaning parts, inspect them for excessive wear. If the throttle or choke shaft is loose in its bearings, the whole carburetor will have to be replaced. Fortunately, this seldom occurs. Blow out all passages with compressed air. A tire or bicycle pump will do fine. Do not put wires or drills into metering orifices.

To reassemble the carburetor simply follow the disassembly procedure in reverse. Do not immerse the choke diaphragm in the carburetor cleaner. This unit can be checked for leakage by depressing the stem and then holding your finger over the vacuum tube. If the stem moves more than 1/16 inch in 10 seconds, the unit is faulty.

Several adjustments need to be made. Dry-float setting, choke-diaphragm pull or qualify, dechoke, curb and fast-idle settings, accelerator-pump, fast-idle-cam phasing and dashpot setting are all needed. The accompanying pictures should help.

13/Set dry float by measuring float protruding above the casting with light pressure on the float tab as shown. Consult repair kit, manufacturer's service manual or the Holley Spec Manual for the correct dimension. Adjustment is made by bending tab.

14/Hook a vacuum source (use your engine) to the choke diaphragm and check qualify dimension with a drill as shown, applying a slight closing force to the plate. With the Model 1940, the diaphragm can be moved to the full pull position by pushing in on the stem with your finger. Specification is in the repair kit and Holley Spec Manual. Adjust by bending the choke-diaphragm link.

16/Rough curb-idle adjustment is set by bottom screwdriver (A) to two full turns from having the throttle plate seated in the bore. Rough fast-idle is set by making 3-1/2 turns after initial contact of the screw with the top step of the fast-idle cam (top screwdriver B). Set both to recommended specs when the carburetor is on the engine.

18/Set fast-idle-adjustment screw (arrow) against top step of the cam as shown. Slight pressure on the plate should yield a 3/32-inch clearance between casting and plate. This cam-phasing adjustment is made by bending the choke link (arrow).

15/Dechoke specification is measured between choke plate and casting with the throttle wide open. Adjust by bending tab (arrow).

17/Adjust pump by bending link, with link in the middle slot 2, so bottom edge of lever travels a total of 0.4 inch. Position 1 or inside slot decreases pump capacity; outside position 3 increases capacity.

Model 1940 Exploded View Identification for drawing on page 190

1	Choke plate
2	Choke shaft & lever assembly
3	Choke thermostat lever
4	Choke control lever
5	Choke plate screw
6	Choke bracket swivel screw
7	Air horn to main body screw & LW
8	Throttle adjusting screw
9	Throttle body to main body screw & LW
10	Choke thermostat lever screw
11	Choke thermostat cover clamp screw
12	Fast idle cam screw
13	Fast idle adjusting screw
14	Pump rod clamp screw
15	Choke diaphragm cover screw & LW
16	Throttle plate screw
17	Rubber plug
18	Fuel inlet seat gasket
19	Flange gasket
20	Spark valve gasket

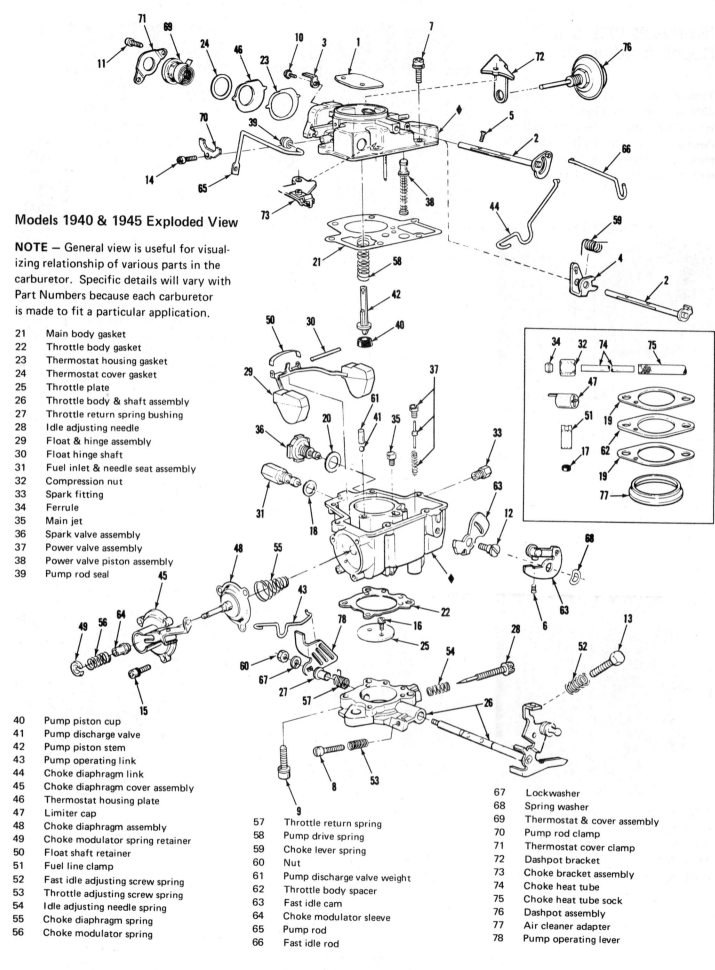

Models 1940 & 1945 Exploded View

NOTE — General view is useful for visualizing relationship of various parts in the carburetor. Specific details will vary with Part Numbers because each carburetor is made to fit a particular application.

21 Main body gasket
22 Throttle body gasket
23 Thermostat housing gasket
24 Thermostat cover gasket
25 Throttle plate
26 Throttle body & shaft assembly
27 Throttle return spring bushing
28 Idle adjusting needle
29 Float & hinge assembly
30 Float hinge shaft
31 Fuel inlet & needle seat assembly
32 Compression nut
33 Spark fitting
34 Ferrule
35 Main jet
36 Spark valve assembly
37 Power valve assembly
38 Power valve piston assembly
39 Pump rod seal

40 Pump piston cup
41 Pump discharge valve
42 Pump piston stem
43 Pump operating link
44 Choke diaphragm link
45 Choke diaphragm cover assembly
46 Thermostat housing plate
47 Limiter cap
48 Choke diaphragm assembly
49 Choke modulator spring retainer
50 Float shaft retainer
51 Fuel line clamp
52 Fast idle adjusting screw spring
53 Throttle adjusting screw spring
54 Idle adjusting needle spring
55 Choke diaphragm spring
56 Choke modulator spring

57 Throttle return spring
58 Pump drive spring
59 Choke lever spring
60 Nut
61 Pump discharge valve weight
62 Throttle body spacer
63 Fast idle cam
64 Choke modulator sleeve
65 Pump rod
66 Fast idle rod

67 Lockwasher
68 Spring washer
69 Thermostat & cover assembly
70 Pump rod clamp
71 Thermostat cover clamp
72 Dashpot bracket
73 Choke bracket assembly
74 Choke heat tube
75 Choke heat tube sock
76 Dashpot assembly
77 Air cleaner adapter
78 Pump operating lever

Models 2210 & 2245
Repair & Adjustment

These two models are very similar. Model 2210 was used for the system descriptions; 2245 is used for these repair photos. Place the carburetor on a stand (Holley's plastic carburetor legs or 5/16 bolts and nuts) to prevent throttle-plate damage.

1/Remove nut and lock-washer from the accelerator-pump shaft (wrench). Remove rocker arm from the shaft. Remove pump rod from rocker arm on one end and from throttle lever on the other. Remove nut retaining choke lever to choke shaft (arrow). Remove lever and link to the fast-idle-cam. Watch fast-idle-cam retainer clip (arrow), they're easy to lose. Remove clip and cam.

3/Remove the clip holding bowl-vent lever and slide lever from the pivot pin, being careful not to lose spring mounted inside of lever. Remove eight screws holding air horn to main body. A slight rap with a plastic hammer or your screwdriver handle may be necessary to separate these castings.

5/Now for the air horn. Push up on the accelerator-pump plunger while pushing in on the shaft to allow disassembly of the pump plunger, spring and retainer. Rotate accelerator-pump shaft until the short end is in its most vertical position and slide shaft from casting. Now remove fuel-inlet fitting and gasket from air horn. With the air horn inverted, remove fuel-inlet baffle from casting by removing retaining screw. Remove Nylon float hinge pin and float. Turn air horn over and let fuel-inlet valve drop out. With a wide-blade screwdriver, remove fuel-inlet seat and gasket. Remove air-horn gasket and discard. If you have to scrape off a portion of this gasket be careful not to damage the sealing surface. Photo shows air horn completely disassembled except for power-valve piston.

2/Remove two screws holding choke-diaphragm bracket to air horn. Disconnect vacuum hose at throttle body to allow removal of the diaphragm assembly, including link to choke-operating lever.

4/Lift air horn straight up, being careful not to damage main-well tubes pressed into upper casting. Tighten attachment screws securely upon reassembly. Go around a second time just to make sure. Use 30 in. lb. if you have a torque wrench.

6/If power-valve piston is binding or sticking, it will have to be removed. Because this requires removing staked material from around retainer and restaking upon reassembly, don't do it if you don't have to. A screwdriver and a few raps with a hammer can accomplish the restaking operation. Arrows indicate staking.

7/Loosen screws holding vent-valve cover and remove cover, valve and spring. Remember, you can't immerse this valve into strong carburetor cleaner. Check the choke plate and shaft for binding or damage. If either exist, remove the choke assembly. Don't do this if you don't have to because the staking must be removed from the attachment screws. This requires restaking the screws upon reassembly, supporting the choke shaft as you do to prevent bending.

9/With assembly inverted, remove five screws holding throttle body to main body.

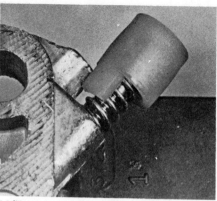

11/Turn each idle-mixture screw clockwise (lean direction) until limiter caps reach their stops. Photo shows idle limiter in lean position. Pry off idle-limiter caps with screwdriver or pliers, being careful not to turn or bend the idle screws while doing this. Turn the screws in with your fingers until they are seated, counting the turns as you go. Record this information and use it on reassembly if the kit contains new limiter caps. Now remove the screws.

8/Take remaining portion of the carburetor from the stand and turn it upside down. Be sure to catch the accelerator-pump discharge valve as it drops out.

10/Now begin disassembly of the main body. Make sure you remove all of the main-to-throttle-body gasket. Use a wide-blade screwdriver to remove power valve from bottom of fuel bowl, being careful not to bend power-valve stem. Remove valve-seat body, valve and spring. Using the same screwdriver, remove main-metering jets, being careful not to destroy the jet slots. Remove fast-idle-cam retainer and the cam. Main body is now completely disassembled and ready for cleaning.

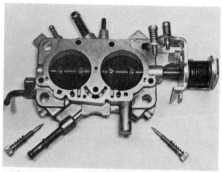

12/It usually is not necessary to remove the throttle plates from the throttle body unless there is damage or wear. If you remove the plates, the staking material must be removed from the attachment screws. These screws must be restaked upon reassembly to prevent them from loosening and causing engine damage. The carburetor is now ready for cleaning by whichever method you've chosen. Photo shows throttle-body assembly with idle-mixture screws removed.

Remember, do not immerse any of the plastic or rubber parts into the stronger commercial type cleaner—only the metal parts. Clean all of the surfaces thoroughly and blow out all of the passages with compressed air.

Before reassembly, inspect all of the parts to be reused for undue wear or damage. Pay particular attention to the choke and throttle shafts and the float. In most cases the parts covered in the kit such as inlet and power valves, the pump cup and all the gaskets are the only ones that need replacing.

In reassembling the carburetor, simply follow the disassembly process in reverse. This carburetor is a relatively easy one to assemble, and the accompanying photographs will help. Screws attaching the air horn and throttle body to the main body require about 30 in. lb. of torque for good retention. If you don't have a torque wrench turn them down until they are snug.

Photographs of all the adjustments are also shown. These include float setting and drop, rough curb and fast-idle speed, choke vacuum-kick or qualify, dechoke or unloader setting, accelerator pump and vent-valve clearance.

14/Float drop is adjusted by bending tab (arrow) so that bottom surface of the float is parallel with the ground.

15/Normal or curb-idle adjustment. You'll want to adjust engine RPM later, but three turns from having the throttle plates seated in the bore is a good starting point.

17/Adjust screw for 0.025-inch clearance between throttle plate and bore. This is a good fast-idle-speed setting. If you have a tachometer, speed should be about 1700 RPM with the screw in the same position and the engine warm.

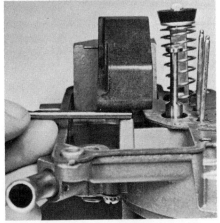

13/Measure the dry float setting between the float and casting with the assembly inverted as shown. Specification is in the repair kit along with a gage. The Holley Spec Manual is another good source. Make the adjustment by bending the tab at the inlet-valve end, being careful not to mar the contact surface.

16/With the fast-idle screw set on the high step of the cam . . .

18/Choke qualify is measured between choke plate and casting at the high side. Use a vacuum source (use your engine) to activate the diaphragm. Measure clearance with a slight pressure on the choke plate as shown. Specification is in repair kit instruction sheet or Holley Spec Manual. Adjust by bending the link.

19/Dechoke spec is checked between plate and casting with throttle held in the wide-open position. Again, see the repair kit or Holley Spec Manual for the proper dimension. Adjust by bending the tang on the throttle lever (arrow).

21/Vent valve should have 0.015-inch clearance over stem with throttle at normal or curb idle position. Adjust by bending tang (arrow).

20/Bend the pump rod until you get a 5/8 pump-stem-to-casting dimension as shown with the rod in the inside or pump-lever position 1. Outside position 2 decreases pump capacity.

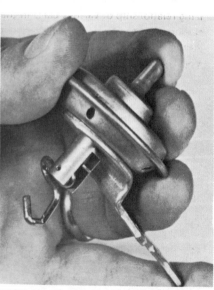

22/Choke diaphragm can be checked by pushing the stem in and holding your finger over the end of the vacuum tube. If the stem moves more than 1/16 inch in ten seconds, replace the assembly.

Models 2210 & 2245 Exploded View

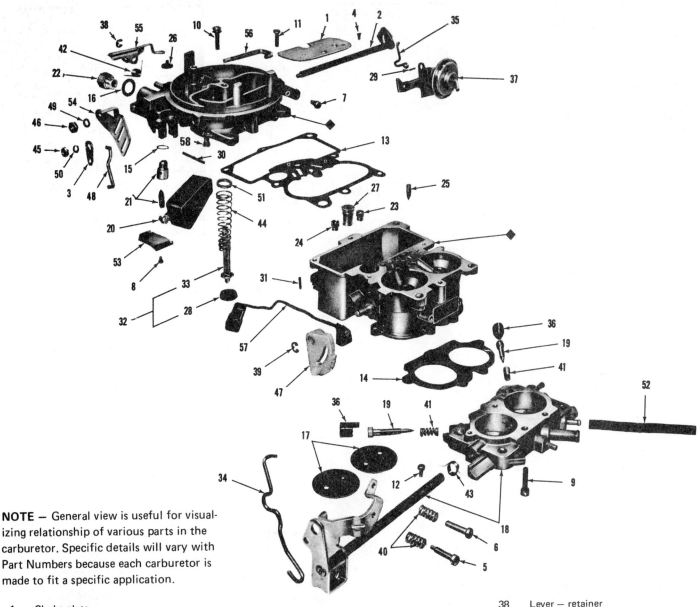

NOTE — General view is useful for visual-izing relationship of various parts in the carburetor. Specific details will vary with Part Numbers because each carburetor is made to fit a specific application.

1	Choke plate	19	Idle adjusting needle	38	Lever — retainer
2	Choke shaft & lever assembly	20	Float assembly	39	Fast idle cam retainer
3	Fast idle lever	21	Fuel inlet needle & seat assy	40	Spring
4	Choke plate screw	22	Fuel inlet fitting	41	Idle needle spring
5	Screw	23	Main jet (choke side)	42	Vent valve lever spring
6	Screw	24	Main jet (throttle side)	43	Throttle lever spring
7	Bracket retainer screw	25	Pump discharge valve	44	Pump drive spring
8	Fuel bowl baffle screw	26	Air vent valve	45	Fast idle lever nut
9	Throttle body to main body screw & LW	27	Power valve assembly	46	Pump lever nut
10	Air horn to main body screw & LW	28	Pump cup	47	Fast idle cam
11	Air horn to main body screw & LW	29	Choke diaphragm link pin	48	Fast idle rod
12	Throttle plate screw & LW	30	Float hinge pin	49	Lockwasher
13	Main body gasket	31	Pin	50	Lockwasher
14	Throttle body gasket	32	Accelerator pump assembly	51	Washer (pump)
15	Fuel inlet needle seat gasket	33	Pump stem assembly	52	Choke vacuum hose
16	Fuel inlet fitting gasket	34	Pump link	53	Fuel bowl baffle
17	Throttle plate	35	Choke diaphragm link	54	Pump lever
18	Throttle body & shaft assy	36	Limiter — carb. idle adj. needle	55	Lever — air vent valve
		37	Choke diaphragm assy complete	56	Pump lever shaft
				57	Stop & cable assembly
				58	Power valve piston assy.

1/Model 5210 on the left is used on GM; 5200 on the right is used on Ford. Both carburetors are very similar in design. Most are used on 4-cylinder engines. Read the repair preparation section on page 162 before starting. Mount the carburetor on a stand to protect the throttle plates from damage. Use the previously described 5/16 nuts and bolts or Holley's plastic carburetor legs. It's a good idea to loosen the fuel-inlet fitting while the carburetor is still intact. Remove the fuel-inlet fitting, gasket, filter and the spring. A new gasket and filter are in the repair kit. The 5200 has a non-replaceable plastic filter.

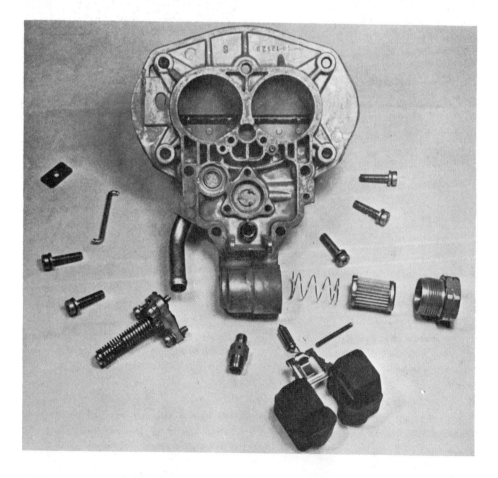

2/Remove retainers from both ends of the choke-operating link and free this link from both levers. Remove five screws holding the air horn to the main body and lift the air horn straight up. A slight rap from a plastic hammer or the handle of your screwdriver may be required to separate the two castings. Slide the choke-operating link through the slot in the air horn and turn the casting to the side, letting the little plastic guide slide out. Remove the float hinge pin with a small drift punch. Lift out the float and inlet valve. Remove the three screws that hold the power valve and remove this assembly. Now it will be easier to remove the fuel-inlet seat and gasket. Photo shows how far you usually have to go with the air horn. If the choke plates or shaft are damaged or binding you will want to remove them also. First, file the staking from the threaded end of the screws and remove these screws with great care. Now slide the plates from the slotted shaft and the shaft from the air horn. When you reassemble the choke plates, hold the air horn up to the light to make sure the plates are properly seated before tightening the attachment screws. The screws will have to be restaked with a prick punch or small chisel. Support the shaft while you do this.

3/The choke assembly. Before doing anything else, note the relationship between the mark on the bi-metal housing and the marks on choke casting. You'll want to restore this bi-metal setting on reassembly. As you face the choke housing, clockwise rotation causes a richer setting.

5/Choke and diaphragm assembly. Replace O-ring 1 with a new one. This passage provides manifold vacuum to the choke diaphragm. Note the relationship between the fast-idle cam, choke lever and spring. Tang 2 is bent for dechoke adjustment. Fast-idle screw is 3. Remove the O-ring from the vacuum passage. Remove three screws holding choke diaphragm cover. This allows you to remove the cover, spring and diaphragm and stem assembly. Normally, this is as far as you will have to go.

7/Choke diaphragm assembly. Check diaphragm for holes or cracks.

4/Next, remove the hot-water-housing retaining screw, washer, housing and the gasket. This may require a rap from your screwdriver handle. Next, remove the three screws and retainer holding the bi-metal housing and gasket. Do not remove the bi-metal from the housing. Do not place the plastic gasket into a harsh cleaner. Remove three screws holding choke casting to main body, allowing you to remove choke casting and assembly. Photo shows hot-water stove and bi-metal assembly removed from the choke casting. Disconnect the fast-idle link and you do this.

6/If you choose to go further, note the relationship of the spring between the fast-idle cam and the choke lever. Refer to the photographs for help. Remove the choke-shaft nut and lock washer. Now the choke lever, fast-idle cam, spring, cup, spacer, and the Teflon lever and the shaft can all be removed. Remove screw retaining the fast-idle-cam lever spacer and washer. Photo shows the choke lever and fast-idle parts disassembled in correct relation to one another.

8/Go next to the main body of the carburetor. Where it applies, remove the two screws holding the anti-dieseling solenoid bracket to the main body. Remove the main air-bleed restrictions and well tubes from the primary and secondary wells. Sometimes they can be a little difficult, so I use a paper clip with a little hook on the end. It slips into the tube so you can pull it. Ordinarily you shouldn't put anything into orifices, but this one isn't critical. Remove primary and secondary idle-retainer plugs and jets on each side of the carburetor.

9/Jet holder (retainer) contains the idle-feed restriction. Idle jets have a number stamped on one of the land surfaces. Remove the main metering jets.

10/Record the numbers as you remove each jet and note where each one came from so you can reinstall it correctly. Jets in this carburetor are different in every location. Some carburetors have idle-feed restrictions 1 with the primary one dyed red and the secondary one a natural-brass color. Bleed restrictions 2 are stamped on top. Well tubes (not shown) are stamped on the bottom. Main jets 3 are stamped on the side. You may have to use a magnifying glass to identify these numbers.

11/Remove the power-valve seat, valve and spring. Do not bend the valve stem. Photo shows main jets 1 and power valve 2 adjacent to their respective tapped holes.

12/Turn the limiter cap (if there is one) as far clockwise as it will go. Remove limiter cap with pliers, taking care not to bend the mixture screw (arrow). Now turn the mixture screw in with your fingers until it reaches its seat, counting turns. This can establish a rough idle setting on reassembly. Remove mixture screw and spring. Remove four screws holding accelerator-pump housing to main body, allowing you to remove the housing, diaphragm and spring. Remove pump-nozzle screw, nozzle, gaskets, and check balls. Normally, this is as far as you would go with the disassembly, especially with the economy method. If you insist on a complete job, be prepared to spend more time and follow the instructions closely. Photo shows accelerator pump and discharge nozzle disassembled. Discharge check and nozzle are installed in hole (arrow). Some list numbers have 2 discharge-check balls.

13/Next, remove the locking retainer (arrow) from the primary throttle shaft nut by bending away the metal tabs.

14/Remove retaining nut 1, and all mechanism attached to the primary throttle shaft can be removed, which includes the primary throttle lever 2, secondary operating lever 3, idle lever 4, return spring 5, and several spacers.

Photo shows relation of all primary and secondary throttle levers and their return springs. If you decide to remove these, this photograph is worth more than the proverbial 1000 words. Unlike the Model 5200, the 5210 secondary-throttle plates must be removed before the secondary throttle lever and spring can be taken off because the level is riveted to the shaft. Unless the throttle bore area is quite dirty or the plates are nicked or the shaft is loose or sticky, don't do it. It is complicated and time consuming and you risk stripping the threads on the throttle shaft. Throttle-plate screws are staked on the threaded side and this staking must be removed with a file before taking out the screws. These screws must be restaked with a small punch upon reassembly to prevent their dropping into the engine. The throttle shafts must be supported during staking to prevent bending. Before torquing down the screws, hold the carburetor up to the light and make sure the plates are seated in the bore when both primary and secondary stop screws are backed off. As you are disassembling the throttle mechanism, pay close attention as to how they relate to each other, particularly the return springs. The accompanying photographs will help here.

15/Note the return-spring positions for primary and secondary.

Now the carburetor is completely disassembled and ready for cleaning. Do the cleaning by either method, but remember: Only metal parts can be immersed into commercial-type carburetor cleaners. Nonmetal parts must be cleaned with a mild solvent such as kerosene or one of the other cleaners mentioned in the repair preparation section. When you've thoroughly cleaned all of the components, blow out the passages with compressed air. If you don't have an air compressor, a hand tire pump will do nicely. Most metering restrictions are removable and easily cleaned. Never stick wires or drills into these restrictions.

Now go through all of the components not included in the repair kit, and hence must be reused. Look for excessive wear. Pay particular attention to the two throttle shafts and the choke shaft. If the bearings are very loose, the complete carburetor will have to be replaced. This is very seldom the case. Check the booster venturi for tightness. A loose fit here could also require carburetor replacement. For reassembly, work your way back through the disassembly procedure. The pictures will be of great assistance.

Several adjustments must be made during assembly. These are dry float setting, bumper-spring adjustment, choke bimetal setting, choke pull-down or qualify, fast-idle-cam phasing, dechoke setting, and rough settings for curb and fast idle. Procedures are demonstrated in the photographs.

16/Choke-qualify dimension. With a light closing force on the choke plate, push diaphragm stem toward diaphragm as far as it will go. With your third hand measure the clearance between the lower edge of the plate and the casting, using a drill (arrow). Correct dimension is listed in repair kit or Holley Spec Manual.

17/Choke qualify is adjusted as shown. Turning in screw gives a smaller or richer qualifying dimension. Turning out has the opposite effect. Some Model 5200's have a slotted adjustment screw inside the removable cap.

18/Dry float setting is measured between the float lung and casting as shown—without gasket and with the air horn inverted. Adjust by bending float tang contacting the inlet valve. Be careful not to mar the contact surface. Dimension is in the repair kit or Holley Spec Manual. Vehicle service manuals are another source.

19/Make sure bi-metal loop fits over the tang upon reassembly. Rotate bi-metal housing to make sure choke plate moves.

20/Fast-idle-cam phasing. With a 5/32-inch drill between choke plate and casting as shown, there should be 0.010 to 0.030-inch clearance between choke-lever tang (pencil points to it) and the fast-idle cam with the fast-idle screw against the second-highest cam step. Bend tang to adjust. This adjustment assures correct choke and fast-idle-cam relationship.

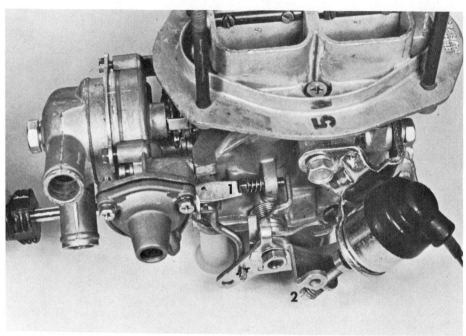

22/Rough normal or curb-idle adjustment 1 is 1-1/2 turns from where the plate is seated in the bore. You'll want to set to correct RPM on the engine. When an anti-dieseling solenoid is used, the idle speed is set with screw 2 with the solenoid activated.

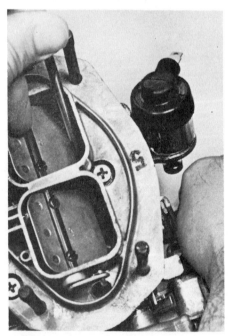

21/Measure the dechoke dimension between the choke plate and casting with the throttle held wide open. Consult repair kit or Holley Spec Manual for correct dimension. Adjustment tang is shown in previous photograph.

23/With the fast-idle adjustment screw against the top step of the cam, a rough fast-idle adjustment is 0.035-inch clearance between throttle plate and casting as shown. Fast-idle screw is shown in the choke casting and assembly photograph.

24/Check to make sure the choke plates seat well in the housing and open and close without sticking or binding. With the air horn off, hold the choke plate and housing up to a light and check for uniform clearance.

25/Secondary idle-speed adjustment 1 should be 1/4 to 1/2 turn from where the throttle plate is seated in the bore.

Model 5200 Exploded View

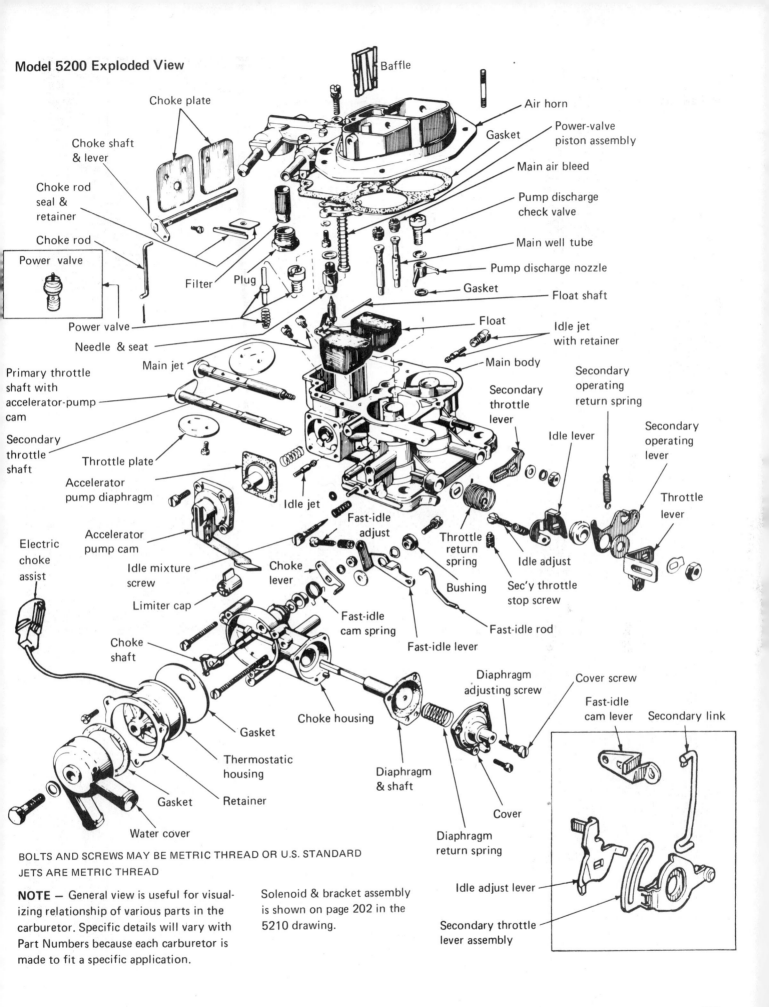

Baffle

Choke plate

Choke shaft & lever

Choke rod seal & retainer

Choke rod

Power valve

Power valve

Needle & seat

Filter

Plug

Air horn

Gasket

Power-valve piston assembly

Main air bleed

Pump discharge check valve

Main well tube

Pump discharge nozzle

Gasket

Float shaft

Float

Idle jet with retainer

Main body

Main jet

Primary throttle shaft with accelerator-pump cam

Secondary throttle shaft

Throttle plate

Accelerator pump diaphragm

Accelerator pump cam

Idle mixture screw

Limiter cap

Electric choke assist

Choke shaft

Gasket

Thermostatic housing

Retainer

Gasket

Water cover

Idle jet

Choke lever

Fast-idle adjust

Fast-idle cam spring

Choke housing

Fast-idle lever

Bushing

Throttle return spring

Fast-idle rod

Sec'y throttle stop screw

Secondary throttle lever

Idle lever

Idle adjust

Secondary operating return spring

Secondary operating lever

Throttle lever

Diaphragm adjusting screw

Cover screw

Fast-idle cam lever

Secondary link

Diaphragm & shaft

Cover

Diaphragm return spring

Idle adjust lever

Secondary throttle lever assembly

BOLTS AND SCREWS MAY BE METRIC THREAD OR U.S. STANDARD
JETS ARE METRIC THREAD

NOTE — General view is useful for visualizing relationship of various parts in the carburetor. Specific details will vary with Part Numbers because each carburetor is made to fit a specific application.

Solenoid & bracket assembly is shown on page 202 in the 5210 drawing.

Model 5210 Exploded View

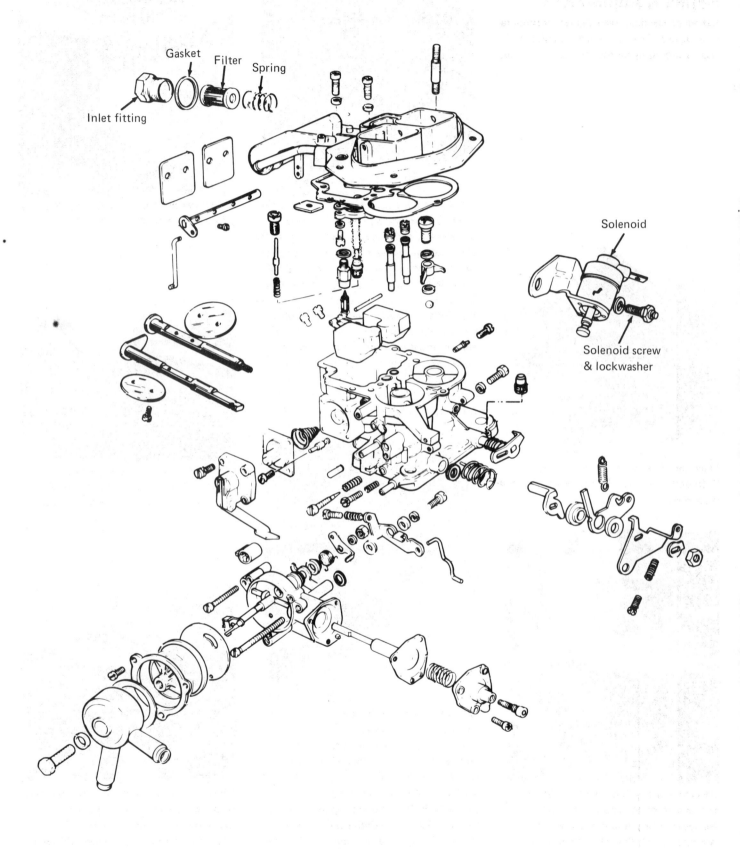

Gasket

Filter

Spring

Inlet fitting

Solenoid

Solenoid screw & lockwasher

ALL BOLTS AND SCREWS ARE U.S. STANDARD, JETS ARE METRIC THREAD
PARTS NOT LABELED ARE ESSENTIALLY THE SAME AS FOR THE 5200 ON PAGE 201.

Model 4360
Repair & Adjustment

Before proceeding, read the repair-preparation section. The accompanying photographs should make the job easier for you.

Place the carburetor on a stand, legs or 5/16 bolts and nuts. This prevents throttle-plate damage.

1/Remove the spring clips from the top and bottom of the choke rod and top of the accelerator-pump operating rods (arrows). Watch the little rascals, they get away easy and are hard to find. Note the accelerator-pump-rod position. You'll want to put it back in the same hole.

2/Remove the two screws 1, 2 holding the choke diaphragm into the main body. Disconnect the choke-diaphragm link from the choke-operating lever and the vacuum hose from the tube in the throttle body. Don't immerse the choke diaphragm in cleaner. Check for leaks by pushing the stem in and then holding your finger over the tube in back. If the stem moves you've got a leak and the diaphragm should be replaced. Next, remove the small C clip 3, holding the choke-operating lever and fast-idle cam onto their common shaft. Be careful, the clip is easy to lose.

3/Before removing the lever and cam, disconnect the small assist spring by lifting the tang out of the hole in the shaft (arrow). Note the spring position. The choke lever and rod are separated by rotating the lever as you remove it. Now you can remove the accelerator-pump-rod from its lever. Note the relationship between the choke lever and fast-idle cam. The torsion spring adds extra closing force for cold starts. Note its relationship to the lever and shaft.

4/Remove the choke rod by rotating it as you pull it through the air horn.

5/Note the position of the choke-rod plastic guide. Choke won't work properly if you reverse the position on reassembly.

6/Remove 10 screws holding the air horn to the main body. Four long ones are across the center of the carburetor. Remove the air horn. It probably will require prying with a screwdriver as shown, and/or a light rap from the handle will help loosen the gasket.

7/Lift the air horn straight up to prevent damaging the power-valve-piston stem. On reassembly, be very careful not to bend the power-valve stem. Get the pump piston started correctly in its bore.

8/Remove accelerator-pump lever by taking out the screw-fulcrum. Unhook the lever from the pump rod, freeing all pump components. Replace the pump cup with a new one on reassembly. Remove the rubber boot from the air horn. Replace it on reassembly.

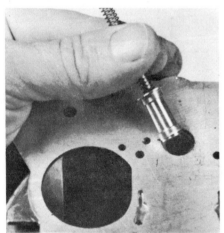

9/If the power-valve-piston is not moving up and down freely, it must be removed and cleaned. Because this requires removing staked material from around the retainer and restaking upon reassembly, don't do it if you don't have to. A screwdriver and a few raps with a hammer can accomplish the restaking operation.

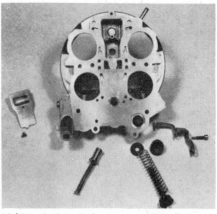

10/Check the choke plate and shaft for binding or damage. If either exist, you'll want to remove the choke assembly. Don't do this unless absolutely necessary because the staking must be removed from the attachment screws. If you perform this operation, be sure to restake the screws upon reassembly, supporting the choke shaft to prevent bending. Here is a disassembled air horn showing the vent baffle, power valve and accelerator pump removed. The air horn is ready for cleaning.

11/Before separating the main and throttle bodies, remove the fuel-inlet plug and fitting with 7/8" and 1" socket wrenches, respectively. Remove the filter, spring and gasket from the inlet cavity. Sintered-metal filter can be cleaned and reused. Remove fuel-inlet-valve assembly with a wide-blade screwdriver and remove the float by lifting the combination hinge and retainer. Turn remaining portion upside down, being careful not to damage primary booster venturis. Now remove six screws holding the main and throttle bodies together. Separate the two. Photo shows main-body components. Remove the main jets from the main body, noting the numbers on the jets so you can put them back in the same position. Primary and secondary main jets are seldom the same size. Remove the power-valve seat, spring and valve. Use a wide-blade screwdriver, taking care not to bend the valve stem.

12/Main body with integral fuel bowl. Components are power valve 1, primary main jets 2, and secondary main jets 3.

13/Close up of main jets and power valve.

14/Remove the limiter caps from the idle-mixture screws, if there are any, with screwdriver or pliers. Turn the mixture screws in with your fingers until they are seated, counting the turns as you go. Record this information for reassembly. Now you can remove the screws and springs. It isn't necessary to remove the throttle plates from the throttle body unless there is damage or wear. Don't do it unless it's necessary—and it's usually not!

15/16/17/If you plan to immerse the throttle body in a strong metal cleaner, you'll have to remove the primary throttle lever and shaft assembly because there is a plastic sleeve under the return spring. If you remove the plates, the staking material must be removed from the attachment screws. These screws must be restaked upon reassembly to prevent them from loosening and causing engine damage. If total disassembly of the throttle body is needed, the photographs show the correct relationship of all levers and springs.

The carburetor is now ready for cleaning. Remember, do not immerse any plastic or rubber parts into the stronger commercial type cleaners—only the metal parts! Clean all surfaces thoroughly and blow out all of the passages with compressed air.

18/Before reassembly inspect all of the parts to be reused for undue wear or damage. Pay particular attention to the choke and throttle shafts and the float. In most cases the kit parts—such as inlet and power valves, the pump cup and all the gaskets—are the *only* ones that need replacing.

In reassembling the carburetor simply reverse the disassembly process. Screws attaching the air horn and throttle body to the main body require about 30 in. lbs. of torque for good retention. If you don't have a torque wrench, turn them down snugly.

Remember to check the float setting before mounting the air horn. Float can be set dry by inverting the main body and holding the retainer while the float drops free. Dimension at point shown should be 1/8-inch.

19/Idle-speed adjustment screw. Back off screw until throttle plates seat in the bores. Then rotate screw 1-1/2 turns clockwise for a beginning idle set. Reset idle speed to specs after carburetor is installed.

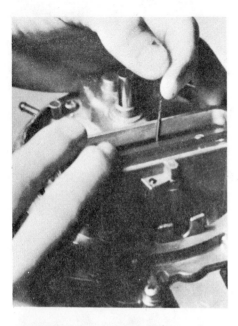

20/21/Choke-qualify set. Apply a vacuum source to the choke diaphragm. Apply a slight closing force to the choke lever with one hand while measuring the clearance at the low side of the plate with the other. Consult the Holley Parts and Specs Manual for the correct dimension. 5/32" will be close if you have to guess. Adjust by bending the link (arrow). Even though there is a de-choke provision, adjustment is unnecessary because the mechanical secondaries provide more than adequate unloading.

Part Identification for Exploded View on next page:

1 Choke Plate	42 Throttle Lever Ball
2 Choke Shaft & Lever Assembly	43 Pump Cup
3 Choke Control Lever	44 Choke Rod Seal
4 Fast Idle Cam Lever	45 Pump Stem Seal
5 Dechoke Lever	46 Accelerating Pump Assy.
6 Kill Idle Adjusting Screw	47 Choke Diaphragm Link
7 Air Horn to Main Body Screw Short	48 Choke Diaphragm Assy.
8 Solenoid Bracket Screw & L.W.	49 Choke Control Lever Ret.
9 Fast Idle Adjusting Screw	50 Pump Lever Stud
10 Choke Diaphragm Bracket Screw	51 Power Valve Spring
11 Fuel Bowl Baffle Screw	52 Kill Idle Screw Spring
12 Choke Plate Screw	53 Idle Needle Spring
13 Air Horn to Main Body Screw Long	54 Fuel Inlet Filter Spring
14 Throttle Body to Main Body Screw & L.W.	55 Fast Idle Screw Spring
15 Throttle Plate Screw Pri.	56 Drive Spring
16 Throttle Plate Screw Sec.	57 Fast Idle Cam Lever Return Spring
17 Dechoke Lever Screw & L.W.	58 Choke Control Lever Spring
18 TEE Plug	59 Throttle Lever Ball Nut
19 Fuel Inlet Plug	60 Fast Idle Cam Assy.
20 Power Brake Plug	61 Choke Rod
21 Fuel Inlet Filter Gasket	62 Secondary Connecting Rod
22 Fuel Inlet Fitting Gasket	63 Accelerating Pump Rod
23 Main Body Gasket	64 Throttle Lever Ball L.W.
24 Throttle Body Gasket	65 Connecting Rod Washer
25 Fuel Inlet Plug Gasket	66 Dechoke Lever Retaining W.
26 Fuel Valve Seat Gasket	67 Spring Perch Washer
27 Flange Gasket	68 Connecting Rod Retainer
28 Throttle Plate Pri.	69 Choke Rod Retainer
29 Throttle Plate Sec.	70 Pump Rod Retainer
30 Throttle Body & Shaft Assy.	71 Solenoid Bracket
31 Idle Adjusting Needle	72 Choke Vacuum Hose
32 Float & Hinge Assy.	73 Choke Vacuum Hose
33 Float Hinge Shaft & Retainer	74 Fuel Bowl Baffle
34 Fuel Inlet Valve Assy.	75 Fuel Inlet Filter
35 Fuel Inlet Fitting	76 Accelerating Pump Lever
36 TEE Connector	77 Solenoid Idle Stop
37 Main Jet Primary	78 Solenoid Nut
38 Main Jet Secondary	Parts not shown on illustration
39 Power Valve Assy.	P.C.V. Tube Plug
40 Power Valve Needle Seat	Throttle Lever Ball
41 Power Valve Needle	Throttle Lever Ball L.W.
	Throttle Lever Ball Nut
	Trans Kick-Down Stud
	Trans Kick-Down Nut

22/Rough fast-idle adjustment. With the fast-idle-adjustment screw (arrow) set on the top step of the cam, adjust so a 0.025" drill (No. 72) will fit between the throttle plate and bore as shown. If you don't have a drill, just estimate and reset to specification on the vehicle. You'll probably want to do this anyway. If you don't have the spec, about 1700 RPM with the engine warm and the vehicle in neutral is a good starting place.

Model 4360 Exploded View

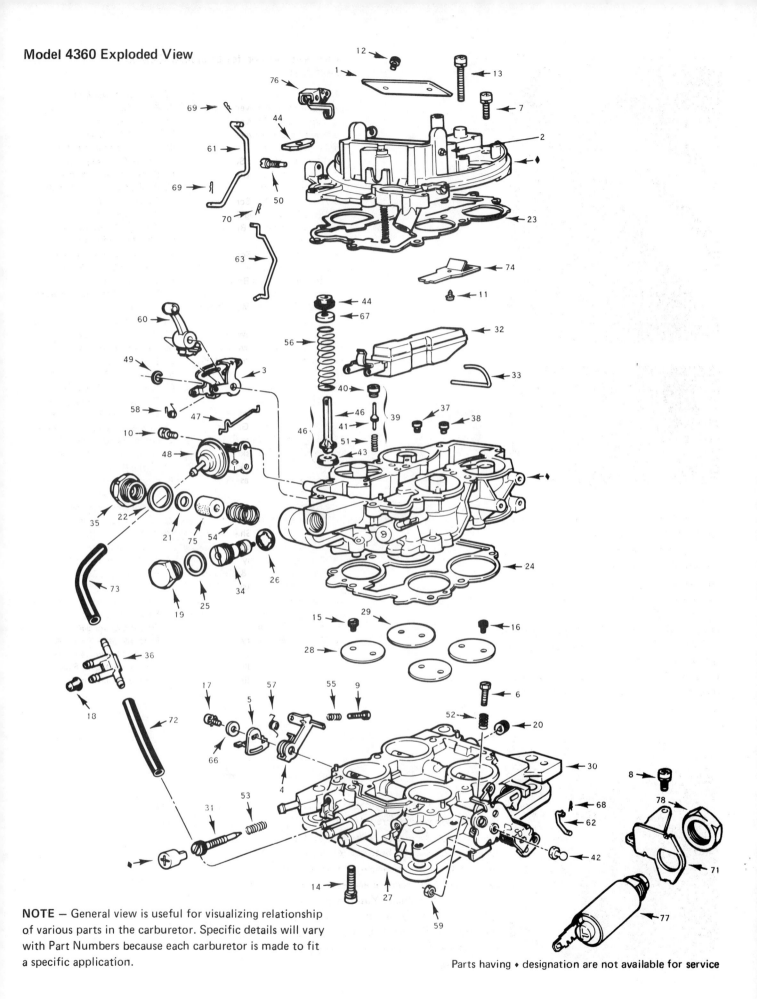

NOTE — General view is useful for visualizing relationship of various parts in the carburetor. Specific details will vary with Part Numbers because each carburetor is made to fit a specific application.

Parts having ◆ designation are **not available for service**

Race bowls with individual fuel inlets require plumbing. It takes effort to gather the various pieces to mate with the line from the fuel pump. Holley's universal fuel line kit 52R-187A plumbs all 4500's, double-pumpers, 3310's and 4165's— and any unit converted to race bowls. 3/8-inch copper tubing is silver soldered into the junction block, which also has a 1/8-inch NPT tap for a fuel-pressure-gage takeoff.

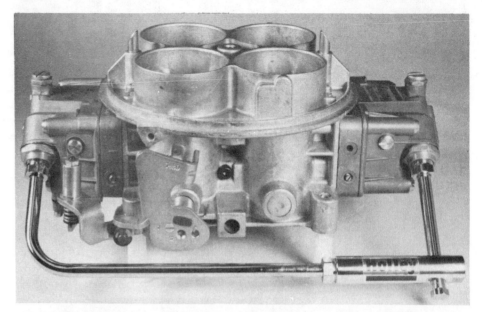

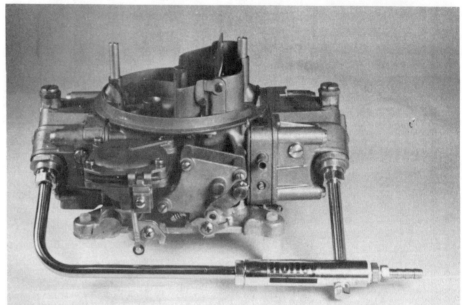

Open throttles extend below the throttle body. Handle the carburetor with care when it is off of the manifold. Avoid letting your friends play with the carburetor unnecessarily! Damage inevitably occurs when some unknowing person holds the throttles open—then sets the carburetor down hard onto a workbench or other surface. Chances are good that you won't be able to run the carburetor until you've purchased new throttle plates—at your expense.

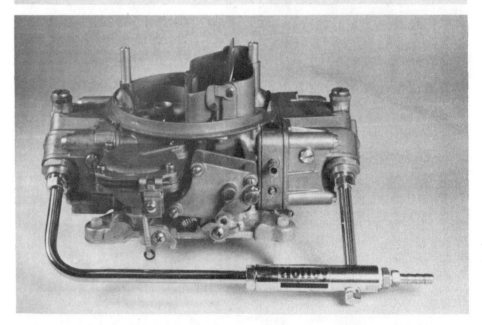